PINEAPPLE
Production & Processing

PINEAPPLE
Production & Processing

By
Desh Beer Singh

2012
DAYA PUBLISHING HOUSE®
Delhi - 110 002

© 2012, DESH BEER SINGH (b. –   )
ISBN 9789351241454

Published by	:	**Daya Publishing House®** **A Division of** **Astral International Pvt. Ltd.** **– ISO 9001:2008 Certified Company –** 4760-61/23, Ansari Road, Darya Ganj, New Delhi - 110 002 Phone: 23245578, 23244987 Fax: (011) 23260116 e-mail : dayabooks@vsnl.com website : www.dayabooks.com
Laser Typesetting	:	**Classic Computer Services** Delhi - 110 035
Printed at	:	**Chawla Offset Printers** Delhi - 110 052

PRINTED IN INDIA

Preface

Pineapple is the third most important tropical fruit in world production after banana and citrus. Seventy percent of the pineapple production in the world is consumed as fresh fruit in the country of origin. It is highly nutritious with exquisite flavour well suits for processing into various value added products. The worldwide production has developed since the early 1500 s when pineapple was first taken to Europe and then distributed throughout the worlds tropics. Being a hardy plant pineapple is grown with least care in humid regions.

International trade is dominated by few multinational companies that have developed the infrastructure to process and market pineapple.

Lack of scientific production management/high cost of production and poor infrastructure for post harvest management/technology and processing for different value added products are the main reasons for its poor productivity and high cost of production. Unsystematic spacing, continuous ratooning, inadequate manuring and other crop husbandry practices contribute to its poor yields.

This book is an attempt to put in one place a synopsis of worldwide research findings and commercial practices. The book integrates botany, taxonomy, breeding, physiology, pathology, on farm handling, safe harvesting, storage and processing with commercial and scientific practices. Efforts has been made in this book on pineapple production and processing and to consolidate research information contained in sufficient and relevant to subject, covering all aspects of pineapple production and processing in a simple, straight and comprehensive manner.

Successful commercial production and processing requires not only integration of these findings but their management in different biotic and abiotic environment and different market needs.

It is hoped that information provided in this will book be quite useful for farmers, extension personnel, researchers, students, and teachers and all concerned with pineapple cultivation and processing.

Help, support and suggestions by my colleagues at Central Agricultural Research Institute, Port Blair and Central Institute of Post Harvest Engineering and Technology, Ludhiana are duly acknowledged.

Desh Beer Singh

Contents

Chapter 1
Introduction, History Origin, and Distribution

1.1 Introduction and History

The pineapple (*Ananas comosus*) originated in the America and became widely distributed in the tropics only in the mid sixteenth century when the cultivated species *Ananas comosus* became recognized for the excellent qualities of its fruits. Before discovery of pineapple fruits by Cristobal Colon on (Christopher Columbus) Nov. 1493 (Morison 1963) pineapple was in an Indian village in an island of Lesser Antilles. The fruit was already a stable component of the vegetative crop complex and in the diet of Native Americans in the low land tropics (Laufer, 1929). Early reports indicate that domesticated pineapple was already very widely distributed in the Americas (Orinoco,

Amazon, Coastal Brazil around Rio de Janeiro) and the Caribbean prior to arrival of Columbus (Collins, 1960). The name pineapple is derived from Spanish name 'Pina' given to the plant, based on appearance of its fruits, which resembles a pine cone. The name 'Ananas' which later became the generic name, is derived from Tupi Indian name 'Nane'.The names of pineapple as 'nanas' and 'ananas' were broadly used throughout South America and the Caribbean. Early European explorers observed a high degree of domestication of this crop. The Amerindians easily distinguished landraces from the wild types and had developed a thorough knowledge of the crop, including production techniques. Variation in fruit yield and quality were found in adapted land races (*e.g.* the Andeans 'Perolera' and 'Manzana').

In addition to the fresh fruit pineapple was used for the preparation of alcoholic beverages (Pineapple wine, chicha and guarapo), for the production of fiber, and for medicinal purposes (emmenagogue, abortifacient, antiamoebic and vermifuge, stoma chic disorders and for the poisoning of arrowheads), mostly related to the proteolytic enzyme bromelain of the pineapple (Leal and Coppens d' Eeckenbrugge, 1996). The Native Americans also domesticated the Curragua, a smooth leaved genotype with a higher yield of long and strong fibres, and used it for making nautical and fishing lines, fishing nets, hammocks and loincloths (Leal and Amaya, 1991). There is still a small traditional industry based on pineapple fiber in Brazil (Leme and Marigo, 1993) and even in the Phillipines, where 'pinavcloth' was mentioned as early as 1571 (Collin 1960; Montinola, 1991). From the early 1500s, the pineapple fascinated the Europeans who introduced and grew it in green house cultivation. European green houses expanded during the 18th and 19th centuries and many varieties were imported, mostly from the Antilles, Griffin (1806) described 10 of them and considered most of the other as useless and their cultivation cumbersome.

Present day variety smooth cayenne was introduced from French Guinea by Perrotel in 1819 (Perrotet 1825) with the notable exceptions of the Smooth Cayenne and Queen, most of these early varieties disappeared as commercial cultivation in Europe declined and pineapple fruit was imported from the West Indies.

'Smooth Cayenne' and 'Queen' were taken from Europe to all tropical and subtropical regions. The Spaniards and Portuguese spread other varieties, including ' Singapore Spanish" to Africa and Asia during the great voyage of the 16th and 17th centuries. However, the diversity of these varieties is still negligible compared with variation found in America. Smooth Cayenne is by far the most important variety in world trade.

Interest in pineapple as a commercial crop begins in early nineteenth century. Sizeable industries existed in Australia, in South Africa, Florida, Malaya and elsewhere. The big producing areas today for trade are tropical Australia, Hawaii, Central and South America, South Africa, Malaysia, Philippines and others.

Its pleasant flavour and equisite taste qualifies pineapple as one of the choicest fruit throughout the world. Pigafetta (1519) reported presence of pineapple in Brazil and described it as an exquisite fruit in existence, in his book "Historia Generally Natural de la Indias". Oviedo (1535) stated "there is no other fruits in the whole world to equal them for their beauty of appearance, delicate fragrance and excellent flavour.

Pineapple is a good source of carotene (vitamin A) and ascorbic acid (vitamin C) and is fairly rich in vitamin B and B_2 (Lal and Pruthi, 1955). It also contains phosphorus and minerals like calcium, magnesium, potassium and Iron (Lodh *et al.*, 1972). Besides it is also a source of bromelin, a digestive enzyme (Lodh *et al.*, 1973). It provides adequate roughage to prevent constipation. Its fresh juice has a cooling and refreshing effect, especially in summer. As a appetizer, the juice can be given to patients suffering from liver diseases, nephritis, stomach complaints, heart disease and general weakness. The fruit in addition to being eaten fresh can also be canned and processed in different forms. Pineapple-bran, a dried rag of pulp after pressing of juice is one of the good cattle feed. A very fine fibre is extracted from its leaves for making a light but stiff fabric (Hayes 1960) called Pina Cloth.

1.2 Origin

The geographical distribution of the existing wild species leads to the supposition that the area of pineapple origin is the region bounded by 15-30 ° south latitude and 40-60 west longitude. Bertoni (1919) considered Paraguay as the place of origin of *Ananas comosus*. The cultivated seedless is the result of a natural mutation in the wild seeded pineapple. The name of the genus, *Ananas*, is derived from the Tupi Guarani Indian word "Nana": in country pineapple has apparently originated. It has been suggested that the Tupi-Guarani Indians first selected and cultivated, *A. comosus* at its center of origin and later took it with them on their subsequent migration. Baker and Collins (1939) believed the place of origin to be somewhere in the region including central and southern Brazil, northern Argentina and Paraguay, as the maximum genetic diversity of pineapple is found in this region. They recorded two forms of wild *Ananas comosus*. The wild plants growing in Matto Grosso, Moura Brazil and Rio de Janeiro were more vigorous and fruits were smaller and more palatable than the wild plants growing in other places of Brazil and Trinidard Islands. The modern pineapple is a cultivar, which was domesticated in Pre Columbian times in South America. A fruit size, juiciness, sweetness, and variation in fruit yield and quality improved flavour. Collins (1960) considered that the center of origin was probably in the drainage area of the Panama-Paraguay River, where related seedy species *Ananas bracteatus*, *A. ananassoids*, *A. erectifolius* and *Pseudananas sagenarious* occurred wild and some of these were also cultivated for fibre. The wild species of seeded pineapple are still seen growing naturally in tropical America, and the cultivation of pineapple in Brazil must have been many centuries old (Laufer, 1929). The belief that long continued propagation of the domesticated species by slips and suckers would somehow must have resulted in plants which lost their ability to produce seeds, has become rather generally accepted, and Brazil is recognized as the center of origin of the present day cultivated pineapples.

1.3 Distribution

Tupi-Guarani Indian Tribes are believed to have migrated northwards and westwards taking pineapple along with them and introducing it to other tribes in the new areas. By the dual process of tribal migration and border trading between tribes, the best varieties of pineapple got distributed throughout the tropical America, and the species developed into cultivars in the process. Later, pineapples incited interest and enthusiasm of the early explorers and settlers of America more than any other plant. When these voyagers started going from America to other islands, many of them carried pineapple fruits and sometimes even the entire plant. The pineapple plants, suckers, slips and crowns withstand considerable desiccation and resume growth when planted, thereby, making pineapple easy to establish in new areas. After the discovery of the new world, pineapple spread was very rapid throughout the tropics. They were either introduced into or were reported to be growing in the Old world (St. Helens, 1505: Madagascar, 1548: southern India, 1550: Philippines, 1558: Java, 1599: West Africa, 1602; Formosa, 1650; South Africa, 1660; Mauritius, 1661; Australia, 1839). Pineapple was introduced into Hawaii in the early nineteenth century.

Pineapple is now widely grown throughout the tropics and sub tropics. Their development as an important economic fruit crop in Malaya, Hawaii, Australia, Ceylon, Formosa, India, Java, Philippines, Singapore, South Africa, Sumatra and elsewhere has occurred during the present century. The region between latitude 25° north and 25° south of the equator is regarded as especially favorable for growing pineapples, though cultivation of this tropical fruit also been successful beyond these limits, as is proved by the extensive cultivation in South Africa and Australia.

The total world production of pineapple is about million tones. Among various countries Thailand with 2278566 tones, Phillipines 2016462 tones, Brazil 2676417, India 1308000, Indonesia 2237858, are main producers (FAO,

Figure 1: Pineapple Plants in Fruiting

Figure 2: Ripened Pineapple Fruit

Figure 3: Immature Pineapple Fruit

Figure 4: Pineapple

Figure 5: Pineapple Plantation

2008) (Table 1). The greater part of pineapple grown commercially are used in the processing industry. World pineapple production should exceed 12 million tonnes in 2009. Asia has strengthened its position as the leading production zone year after year. Seventy per cent of pineapple consumption is in production areas. World trade in the fruit consists of three main products : preserves, forming by far the largest proportion (1,000,000 tonnes), pure or concentrated juice (170,000 tonnes) and fresh fruits (500,000 tonnes). Four-fifths of the world preserves and juice market is supplied by Thailand and the Philippines.

Most of the canned products are exported to Europe and North America. Most of the canned products are exported to Europe from South Africa, Kenya, Cuba, Guinea, East Cameroon.

Even though climate prevailing in the large parts of India is ideal for pineapple cultivation, still this fruit crop does not hold any position of importance among the major fruits being cultivated in the country.

In India it is grown mostly in Assam, West Bengal, Tripura, Kerela, Goa, Orissa, Karnataka, and Meghalaya.

Production and Area Under Cultivation

I. In World

Table 1: Pineapple Production, Area Under Cultivation and Yield (t/ha) (FAO, 2008)

Country	Production (Tones)		Area (ha)	
	2007	2008	2007	2008
Thailand	2815275	2278566	94449	93116
Philipines	2016462	2209336	53978	58251
Brazil	2676417	2491974	71886	62142
China	1381901	1402060	66372	70613
India	1308000	1305800	85800	81900
Nigeria	900000	900000	117500	117500
Mexico	671131	685805	15918	16377
Costarica	1968000	1624568	35200	33488
Colombia	434574	436044	10914	10686
Indonesia	2237858	1272761	18957	20802
Venezuela	363075	363075	18507	18507
USA	172500	172500	5700	5700
Kenya	429065	429065	14271	14271
Cota d'ivoire	159668	159668	4562	4562
South Africa	146214	146869	13000	13000
Australia	164732	164732	5134	5134
Dominican Republic	91593	100528	7014	9227

Contd...

Table 1–Contd...

Country	Production (Tones)		Area (ha)	
	2007	2008	2007	2008
Malaysia	316210	319130	10900	10900
Guetemala	230566	230566	7630	7630
Honduras	154000	154000	2800	2800
Cameroon	52000	52000	3650	3650
Cuba	51597	55387	5907	5991
Cambodia	16000	16000	1600	1600

Source: Statistical yearbook 2009. Food & Agriculture Organisation of the United Nations.

Table 2: Production and Area under Pineapple in India 2009–2010.

State/UTs	Pineapple	
	Area	Production
Andaman & Nicobar	0.2	0.6
Andhra Pradesh	0.0	0.0
Arunachal Pradesh	10.9	34.4
Assam	14.2	223.0
Bihar	4.7	125.0
Chandigarh	0.0	0.0
Chhattisgarh	0.0	0.0
Dadar and Nagar Haveli	0.0	0.0
Daman and Diu		
Delhi	0.0	0.0
Goa	0.3	4.5
Gujarat		
Haryana	0.0	0.0
Himachal Pradesh	0.0	0.0

Pineapple: Production and Processing

Table 2–Contd...

State/UTs	Pineapple	
	Area	*Production*
Jammu & Kashmir	0.0	0.0
Jharkhand		
Karnataka	2.8	177.2
Kerala	9.8	80.8
Lakshdweep	0.0	0.0
Madhya Pradesh		
Maharashtra		
Manipur	12.1	103.5
Meghalaya	10.8	106.8
Mizoram	0.4	6.3
Nagaland	8.0	80.1
Orissa	0.7	8.4
Pondicherry	0.0	0.0
Punjab		
Rajasthan	0.0	0.0
Sikkim		
Tamil Nadu	0.5	25.0
Tripura	6.8	117.5
Uttar Pradesh	0.0	0.0
Uttaranchal		
West Bengal	9.6	293.8
Total	91.9	1386.9

Source: Area and production estimated for horticultural crops by NHB 2009–10.

Chapter 2
Botany

2.1 Morphology

Ananas comosus is a herbaceous, perennial, self sterile of the monocotyledonous group, whose terminal inflorescence gives origin to a multiple fruit called as sorose. After maturation of the first fruit, the plant develops new shoots from maxillary buds, so producing new growth axes capable of producing another fruit. Normal fruiting pineapple plant consists of suckers, slips, peduncle and crown in addition to leaves; stem and roots (Figure 6) The same plant thus may give a sequence of various production cycles. Due to reduction in fruit size and uniformity in most commercial plantings, the plants are not allowed to produce more than two to three crops. New plantation may be done with the same lateral shoots of the preceding crop, or with other vegetative propagules, such as the fruit crown, or in many cultivars, slip produced from the peduncle. Vegetative

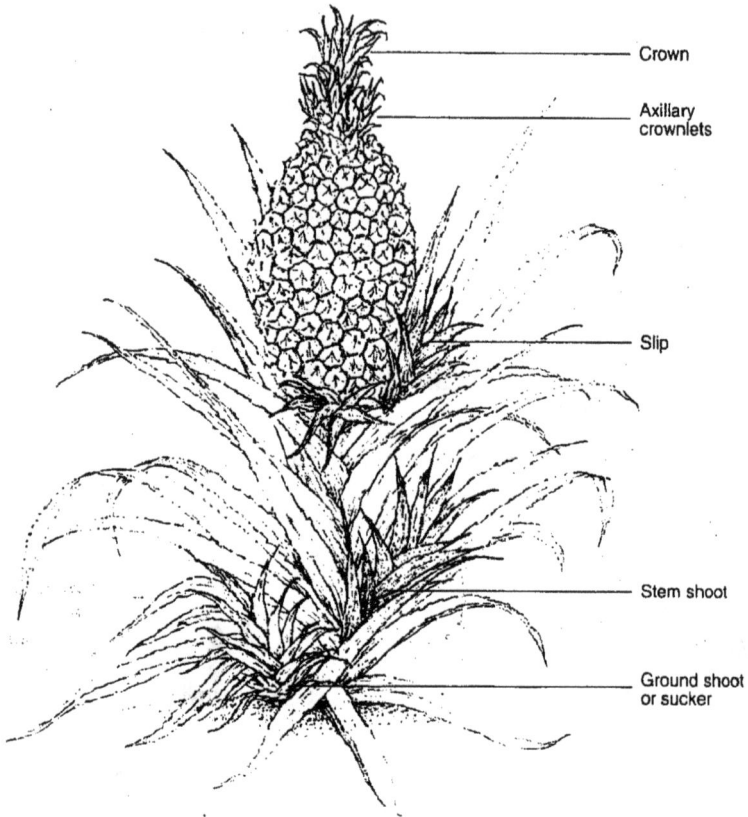

**Figure 6: Main Morphological Structure
of the Pineapple Plant**

reproduction is mostly dominant in wild genotypes, in which in addition to lateral shoots; the crowns and slips contribute to propagation, as they resume proper and rapid growth at maturity. The adult plant is 1-2 m high and 1-2 m wide and it is inscribed in the general shape of a spinning top. The main morphological structures to be distinguished are the stems, the leaves, the peduncle, multiple fruit of the sincere, the crown, shoots and the roots are the main morphological structures to be distinguished.

2.2 Stem

The central stem which leaves the apical meristem, and to which foliage, leaves are attached is given in Figure 6. The pineapple stem is club shaped, short and thick, with a length of 25–50 cm and a width of 2-5 cm at the base and 5-8 cm at the top with short inter nodes. Depending upon the point of origin, the suckers are called ground suckers if they arise from the buds of the underground portion of stem or shoot and slips if they arise from axillary buds of the aerial portion of the stem. Slips are leafy brackets attached below the fruit and are developed from the axillary buds on the peduncle. Peduncle is a slender leaf bearing stalk, supporting fruit and connecting it with stem crown is a miniature plant, consisting of condensed stem and leaves, growing from the apex of the fruit. During fruit development axillary buds in the leaf axils elongate to form lateral branches called shoots. Rarely buds at the base of the peduncle grow out to produce hapas. The basic difference between hapas and suckers is that hapas develop higher of the stem, in transition zone between the stem and peduncle. Its aerial part is straight and erect, while the shape of the earthed part depends on the material used for planting. It is markedly curved when coming from the slip, as the stem of these propagules is comma shaped, less curved when coming from the stem shoot and erect when coming from a crown. Nodes can be visualized by the leaf scars left after stripping the leaves from the stem. Shoot buds, 3-5 mm high and about 5 mm wide, occur in the leaf axils. They are mostly larger because of an increase in the size of their prophyll (the first leaf of the shoot), which encloses it. As striking feature of the pineapple stem is the presence of adventitious roots breaking through the epidermis, and growing flattened and distorted, tightly around the stem, between the leaves.

The stem constitutes a central cylinder or stele and a cortex separated by a thin layer of vascular bundles produced by the dome shaped epical meristem. The dense network of vascular tissue separating cortex and stele consist chiefly of xylem, with very little phloem. In this tissue, areas of non vascular tissue or leaf gaps, are disposed at intervals, allowing leaf trace bundles to pass from the cortex in to the stele. This vascular cylinder is thicker and subrised at the stem base. On the cortical side, a narrow layer of long, thin walled cells bound it. Vascular bundles are very numerous throughout the stem but less so in the cortex than in the stele. The later is mainly constituted of a compact parenchyma with abundant starch. It contains large cells with raphides of calcium oxalate crystals. The cortex is composed of parenchymatous tissues, crossed by the isolated vascular bundles going to the leaves of the adventitious roots originating at the boundry with the central cylinder and of circumferential small bands of vascular tissue lying just above the leaves attachment to the stem. The inner parenchyma of this cortex is also rich in starch and contains raphide cells. Limiting the stem externally is the epidermis, with petat trichomes in the nodal regions.

2.3 Leaves

The leaves of the pineapple are long, thin and tapered towards the tip and are curved upwards in cross section, which appears to be a strengthening adaptation to keep the leaves rigid. The leaves are arranged in a right or left-handed spiral on short stem, forming a rosette. In addition, the fluted nature of the leaves create a channel, which conducts water to the axils in which axillary roots occur. This is considered to be an adaptation to rigid conditions since dew collects in the axils and constitutes an important source of water under low rainfall conditions (Collins, 1968). The number of functional leaves range from the 35 to 60 and there is a bud in every leaf axil. The leaves either have smooth edges with a few spines just below the tip or have spines all along the margin. The tip is elongated, ending in a finer point. Leaves are sessile with clasping base excepting near apex; lamina is shaped like a shallow trough, which conducts water to the base of the plant. The upper leaf surface is green and the lower is silvery white, due to the presence of trichomes. The pineapple leaf possesses a number of xerophytic adaptations. The upper epidermis consists of compact interlocking cells, which are thickened and stratified and coated with a waxy cuticle. Stomata occurs only on the lower surface within longitudinal furrows, reported to be 70 to 85 per mm^2 (Krause 1949); 180 per mm^2 in Cayenne and less for triploid and metalloid; Py and Tisseau 1987 and Perseglove 1972).

The lower and also to a lesser extent, the upper surface bears mushroom shaped hairs (trichomes) which effectively increase the thickness of the boundary layer at the leaf surface and increase air resistance to the diffusion of the gases and loss of water. Trichomes are multicellular hairs with short stalks arranged in furrows; the flattened top of the air filled dead cells spread out and cover ridges and give leaf under surface a white appearance. The trichomes have also been described as 'reflecting harmful rays and preventing excessive heating', and suggested as organs for water absorption. Internally the leaf possesses a layer of water storage tissue below the upper epidermis consisting of columnar, colourless cells, which occupy about half the thickness when fully turgid. These cells collapse and decrease in thickness under dry conditions. Because of this an ordinary system of photosynthesis does not occur and plants use CAM (Crassulacean acid metabolism) in which CO_2 is reduced to organic acids (mainly citric and Malic), which accumulate in leaves at night and are reduced to sugars during day.

Functionally the pineapple leaf is also adapted to survive drought. Joshi *et al.* (1965) showed that stomata opened mostly at night and in the early morning, and suggested that the pineapple is adapted to store CO_2, which is later used when the stomata have closed with the advance of the day. Water loss during the day was estimated at about 0.3-0.5 mg/cm^2/hr. The phyllotaxy of the leaves is 5/13 (number of whorls and number of leaves per spiral or whorl).

2.4 Root System

The pineapple has a restricted root system, which reflects its epiphytic origin.The pineapple has few shallow and a very restricted root system, which reflects the epiphytic origin of the species. Many of the wild species of pineapple are epiphytic and although the cultivated pineapple is terrestrial, its root system is nevertheless very confined. Although it seems likely that the xerophytic nature of the leaves is more related to the restricted root system than reflecting ability to survive severe moisture stress. The roots are shallow and few. In lysimeter study, pineapple roots established to a depth of only 30 cm as opposed to 3 ft or more for other species (Ekern, 1965). Nforzato *et al.* (1968)

stated that in a latosol in Sao Paulo, 95 per cent of the root system occurred in the top 20 cm of soil.

Root growth occurs under moist conditions, but when conditions are dry, it is very much reduced and damage can occur to the superficially placed roots (Black, 1962), but Collins (1968) points out that in low rainfall regions, high humidity and dew formation are necessary for good growth since the roots are able to absorb the axillary water. It is also believed that pineapple roots are associated with mycorhizal fungus (Cobley 1956).

2.5 Floral Biology

2.5.1 Inflorescence, Flower and Fruit

The peduncle and inflorescence develop from the apical meristem about 10-15 months after planting although varies according to cultivars and type of planting material used. The diameter of which is suddenly increased until the initiation of the peduncle (Kern *et al.*, 1936). The stage of inflorescence emergence is called "red heart" because of the five to seven reddish peduncle bracts at its base. These bracts are shorter and narrower than the ordinary leaves. The peduncle elongates after flower formation. Its length varies widely with the botanical varieties or even cultivars. In addition to its bracts, it bears in many cultivars, a variable number of slips (up to a dozen or more), which can be positioned more or less regularly between the stem and the fruit. These slips can be considered as dwarfish 'aborted' fruits with a relatively large crown (Collins, 1960). They may constitute an appreciable source of planting material in extensive cultivation system.

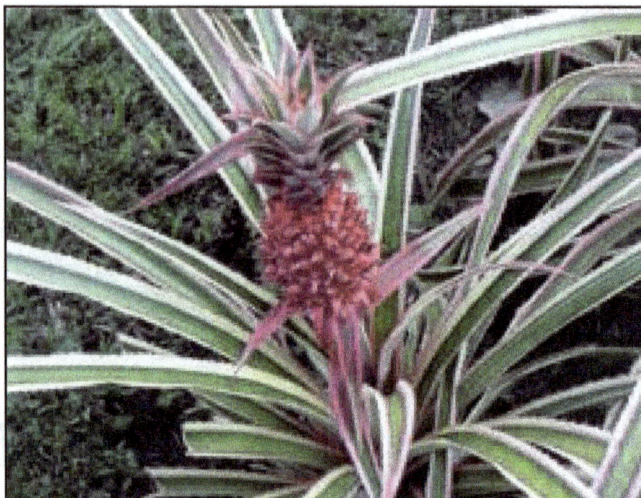

Figure 7: Typical Pineapple Flower

The inflorescence consists of fewer than 50 (in some wild clones) to more than 200 (in some cultivars) individual flowers; it is capped by a crown, composed of numerous short leaves (up to 150) on a short stem. The flowers or individual fruits are disposed around the central axis according to an 8/21 phyllotaxy in large fruited cultivated pineapples and a 5/13 phyllotaxy for small fruited wild pineapples or for young cultivated pineapples flowering prematurely (Kerns *et al.*, 1936). The fibrous axis containing the many vascular bundles that supply the flowers is continuous with the peduncle and with the short stem of the crown. Between the upper most flower and the crown is a transition region with bracts but no flowers. The edible part of the fruit consists chiefly of the ovaries, the basis of the sepals and the bract tissues and the spices of the ovaries (Okimoto, 1948). Anthesis normally takes place within a day. Flowering lasts 10-15 days and occurs in a more or less acropetal succession along the inflorescence axis, but some cultivars flower in a very disorderly manner.

Plant bears total of 100 to 200 reddish purple flowers arranged spirally in the axis. Flowers are hermaphroditic and trimerous, with three sepals, three petals, and six stamens in two whorls of three and one tricarpellate pistil. The anthers are bilobed, introrse and dorsifixed. The hollow, trilobed and trifid style is almost as long as the petals and equal to or longer than the stamens. At anthesis, each stylar canal is an unobstructed open channel from the stigma to the locule directly above the placenta. Petals are ligulate and free, each bearing at its base two slender funnel form scales or, more rarely, lateral folds that overlap the filaments. Petals are white at their base to violet blue at their tip. They are so close together at their outer end that only small insects can enter the flower. This narrow tubular flower and abundant nectar production are particularly adapted to humming bird pollination. Indeed, the three large nectary glands are so productive that nectar often fills the corolla and seeps out. The sepals are deltoid and appear similar to the bracts in the colour and texture. Each flower is surrounded and subtended at its base by a pulpous and thick bract, covered by trichomes, which becomes pointed and papyraceous, at its tip. Parts of three other bracts complete the enclosure of the flower. Bract spineness is correlated with leaf spineness.

The fruit is a syncarp or multiple fruit, found by fusion of fruitlets, produced by each flower. The multiple fruit matures in 5-6 months after flowering. It is formed by extensive thickening of the axis of the inflorescence and by the fusion of the small berry like fruitlets produced by each flower. The sepals and pointed floral bracts, subtending each flower, persist and form hard rind of the fruit lets and become more or less fused in the process with adjacent flowers. In the syncarpic inflorescence resulting from the fusion of the basal part of the flowers are separated by the parenchymatous tissue of the calyx and bract basis. The ovary is inferior, tricarpellate and tricular, with the three septa forming an inverted Y when seen in tangential section of the inflorescence. The placenta and ovules are arranged in two single or double rows. The number of ovules per flower varies with the cultivars, from 16 to 71 (Coppens d' Eeckenbrugge *et al.*, 1993). The occurrence of two types of ovules (unitegmic, orthotropis and bitegmic anatropous) within the same ovary is common and orthotropous ovules are fertilized (Okimoto, 1948; Rao and Wee, 1979; F. Van Miegroet, 1993). Orthotropopus ovules are much less frequent than anatropous ovules and their presence and numbers are a varietal characteristic. Pollen grains are prolate and spheroidal, biconvex, isopolar, and bilaterally symmetrical and diaperturate, with circular to slightly elongated aperatus situated at the poles. The equatorial (36-51 µm) dimensions are variable. The exine is reticulate, and the polar areas show finer reticulation than the rest of the surface (Wee and Rao, 1979).

There is no floral abscission and except for the withering of the style, stamens and petals, the entire blossom develops parthenocarpically into a berry like fruitlet. In the cultivated pineapple, growth from blossoming inflorescence to mature fruits results in a 20-fold increase in weight. The enlargement of the calyx results from continued growth by cellular division in stages up to flowering and cell enlargement, in the later stages. During this size increase, cell walls get thinner. The bract, sepal and ovary tissues are prominent structures in the mature fruit. The large conspicuous in the mature fruit. The large conspicuous bract is fleshy and widened at its base and bends over the flattened calyx surface, covering half of the fruitlet. Its papery tip dries during maturation. Internally, locales get longer but relatively narrower and less conspicuous in the developed fruit because of the expansion of adjacent tissues, especially of the far less than the sepal tissues, especially of the septa. Placentas show some enlargement but far less than the sepal tissues, unless they bear mature seeds. The seeds are approximately 3-5 mm long and 1-2 mm wide, flat on one side and curved on the other, with a pointed end. They contain a hard flinty endosperm and a minute emryo enclosed in a brown to black coat, extremely tough and leatherly and roughened by numerous longitudinal ridges (Miles Thomas and Holmes, 1930). In the mature fruit, the stylar canals get completely closed, first by a mucilaginous plug, soon after anthesisis, and a week or two later by cellular occlusion.

2.5.2 Taxonomy

Pineapple belongs to the order *Bromeliales*, family Bromeliacea, subfamily Bromelioideae. With 2794 species among 56 genera, Luther and Sieff (1998). This is the largest family whose natural distribution is restricted to the new world, with exception of Pitcaurnia Feliciana, which is native to Guinea. The Bromeliaceae have adapted to a very wide range of habitates, ranging from terrestrial to epiphytic, deep shade to full sun, mesic to extremely xeric and sea level to alpine, and from hot and humid species to the cold and dry subtropicals. The Bromeliaceae are set apart from other monocots by the unique, stellate or scale like multicellular hairs and the unusual conduplicate, spiral stigmas (Gilmartin and Brown 1987). They are also characterized by a short term, rosette of narrow stiff leaves, terminal inflorescence, in the form of racemes or panicles, hermaphroditic and actinomorpic trimerous flowers with well differentiated calyx and corolla, six stamens and superior to inferior trilocular ovary with axile placentation and numerous ovules. Fruits are capsules or berries and contain small naked winged or plumose seeds with a reduced endosperm and a small embryo. Most species are epiphytic or saxicolous while some are terrestrial. They are particularly adapted to water economy, based on: (i) rosette structure (ii) ability to absorb water and nutrients through their waxy leaves and aerial roots (iii) ability to store water in specialized equiferous leaf tissue (iv) multicultural trachomes functioning as water valvulae and reflecting radiation (v) a thick cuticle (vi) location of stomatas in furrows, limiting vapor transpiration and (vii) CAM. Their root system is not well developed and functions mostly to anchor the plant.

Pineapple is by far the most important economic plant in the Bromeliaceous. However, in the same Bromelioideae sub family, some *Aechmea* and *Bromelia* species also yield edible fruits, such as *A. bracteata* (Swartz) Grisebach, *A kuntzeana* Mez, *A. longifolia* (Rudge) L.B. Smith and M.A. Spencer, *A. nudicaulis* (l.) Grisbach, *B. antiacantha* Bertoloni, *B. balansae* Mez, *B. chrysantha* Jacquin, *B. karatas* L. B. hemisphaerica Lamarck, *B. niduspuellae* (Andre) Andre ex. Mez, *B. pinguin* L., *B. plumieri* (E. Morren) L.B. Smith and *B. trianae* Mez (Rios and Khan, 1998). The most common are known and consumed locally, under names like cardo or banana–domato (bush banana), pinueles (small pineapple) or karatas, gravata and croata. Many other bromeliads are cultivated as ornamentals, gathered for fibre extraction or used in traditional medicine (Correa, 1952: Purseglove, 1972; Reitz, 1983; Rios and Khan, 1998).

2.5.3 Evolutionary Classification

The first botanical description of cultivated pineapples was by Charles Plumier at the end of the seventeenth century when he collected plants, called karatas and ananas on the islands of Hispaniola. Following the native classification, he created the genus *Bromelia* for the karatas, in honor of the Swedish physician Olaf Bromel, and described the ananas, using polynomials such as *Ananas aculeutus* fructo ovato, *carne albida*. In his Species Plantarum, Linnaeus (1753) designated the pineapple as *Bromelia ananas* and *Bromelia comosa*, While Miller (1754, 1768) maintained the name *Ananas*, with six varieties, all cultivated. In the following classifications of the 18th and 19th centuries, as pineapple was mainly known from attractive large fruited types, these varieties and other cultivars were easily confused with species, which resulted in an overwhelming number of different names (Leal *et al.*, 1998). Lindley (1827) used such names as *Ananassa sativa* for ordinary cultivars, *Ananasa lucida* for smooth leaved cultivars such as Smooth Cayenne (from the variety Ananas (Lucidus) of the eighth edition of the Miller's Gardener's Dictionary, published in 1768), *Ananassa debilis* for a particular cultivars with undulated leaves and *Ananassa bracteata* for a crowned pineapple with long bracts. Schultes and Schultes (1830) returned to the original name *Ananas, A. sativus, A. debilis, A. semmiserratus* (instead of *A. lucida*) and *A. bracteatus*.

Linden (1879) described a Colombian smooth leaved cultivar with the 'piping character under *A. mordilona*. (Morren 1879) gave the first clear description of a distinct pineapple, the yvira, characterized by the long bracts, the absence of a crown and propagation by stolons, which he named *A. macrodontes*. In 1889, both Baker and Andre described a wild pineapple, with a long scape and a small crowned fruit, respectively, under the names of *Acanthostachys ananassoides* and *A. pancheanus*. In contradiction to this profusion of species descriptions, a few authors, such as Bentham and Hooker (cited by Andre 1889) claimed that the genus *Anananas* is monospecific, with multiple wild and cultivated forms. Thus, in his first attempt at simplification, published in the Flora Brasiliensis, Mez (1892) recognized only one species, *Ananas sativus* with 5 botanical varieties. The variety lucidus included the pineapples with smooth leaves and a large fruit, but also the pitte or pitta, a small-fruited pineapple only cultivated for fibre. *A. debilis* was down graded to a second variety debilis only known from European glass houses. The variety bracteatus included *A. bracteatus* (Lindl.) Schult.f, but also *A. macrodentus* Morren. The variety muricatus was made from *A. muricatus* Schult. F.

Many a members of the family in nature are found in tropics and sub tropical regions of the America. Ananas and the closely related monotypic genus Pseudananas are distinguished from other genera of Bromiliaceae by their syncarpous fruits, which in *Ananas* bear terminal crown of the reduced leaves. But pseudananas has no such crown. Both genera have the basic chromosome number n=25, which is common throughout the family, but *Ananas* is typically diploid (2n=50) while pseudananas is tetraploid with 100 chromosomes and could be the result of natural hybridization between a species of Bromelia with 50 chromosomes and a species of Annanasa, followed by chromosome doubling (Collins, 1960). There are 5 species of Ananas- *A. bracteatus, A. fruitzmuelleri, A. comosus, A. erectifolius* and *A. ananassoides* and one species in *Pseudananas, P. sagenarious.*

Both *Bracteatus* and *A. fruitzmuelleri* produce fleshy, edible (though seedy) fruits: *A bracteatus* had in fact been cultivated in the Parana river area (Collins, 1960). Fruits of *A. ananassoids* and *A. erectofolius* become nearly dry at maturity with little flesh. *Ananas erectifolius,* which has long, almost spineless leaves, has been considered a potential fibre-crop. Of the 900 species in the family Bromiliaceae, only their cultivated pineapple belonging to species *A. comosus* (Linn.) Merr is of prime economic importance.

The brief account of pineapple taxonomy is introduced here with a classification showing pineapple lineage.

Kingdom	Vegetable or Plant
Sub Kingdom	Spermatophyta
Class	Angiospermae
Sub Class	Monocotyledons
Order	Farinosae
Family	Bromeliaceae
Genera	*Ananas* and *Pseudananas*

Botanical Key to the Genera and Species (Smith, 1939)

1. Syncarp bearing at maturity a minute, inconspicuous coma of reduced squamiform bract, never producing

slips at its base; plant producing elongate stolons at base; petals bearing appendages in the form of lateral folds....*Pseudananas sagemarius*

1. Syncarp bearing at maturity a conspicuous coma of foliacuous bracts, frequently producing slips at its base; plant not producing stolons; petals bearing 2 infundibuliform scales each......*Ananas*

2. Syncarp well over 15 cm long at maturity, with copious palatable flesh, scape stout and usually short.

   3. Leaf spines at ascending; floral bracts coloured at maturity; petals bearing scales....*A. bracteatus.*

   3. Leaf spines recurved towards the base; floral bracts palegreen at maturity; petals bearing vertical folds....*A.fruitzmuelleri*

   3. Floral bracts relatively inconspicuous and soon exposing, the tops of the ovaries, weakly serrulate or entire; seeds lacking or very rare....*A. comosus*

2. Syncarp 15 cm long or usually much shorter, with scant, unpalatable flesh at maturity; scape elongate or slender

   4. Leaves stiffy, erect, entire, except for a longterminal spine, 35 mm wide....*A. erectifolius*

   4. Leaves recurving, serrate, not over 25 mm wide......*A. ananassoides*

Chapter 3
Varieties and Improvement Breeding

Except *Ananas erectifolius*, all species of pineapple have one way or other some cultivars/varieties. The varieties of *A. comosus* (Linn.) Merr. are of particular interest because of their large number, wide distribution and edible fruits. The horticultural classification of pineapple varieties is of Hume and Miller (1904) is currently followed. They divided cultivated varieties of pineapple into three main groups- (i) Cayenne, (ii) Queen and (iii) Spanish. The pineapple industry of the world is dominated by the cultivar "Smooth Cayenne". which is used both for fresh fruit and for processing. The Smooth Cayenne has been the backbone of the global pineapple industry for more than a century. First collected by Perrotet in 1819 in French Guinea under its local name 'Maipuri (Perrotet, 1825), it rapidly spread to other geographical regions, has adapted well and is known by such other synonymous as 'Kew', Giant Kew, Champaka, and Sarawak. The Smooth Cayenne monopoly is undoubtly due to its high yield, adaptability and good characteristic for canning. Further, highly specialized systems of production and processing protocols have been developed almost exclusively for this cultivar.

Most of the varieties in India may be accommodated into any one of the three groups. Py *et al.* (1987) classified cultivars grown throughout the world into 5 distinct groups.

There is no reliable botanical or horticultural classification of pineapple varieties. Several attempts were made in this direction. Munro (1933) listed 52 varieties grown in England on the basis of flower colour, fruit shape and spine characteristics of the leaves. Only Queen of his listed varieties, is of any importance now. Hume and Miller (1904) designated the pineapple varieties of Florida in to these groups with six varieties. Red Spanish with seven varieties and Cyenne with these varieties (Bertoni 1919).

The important commercial varieties include Cayenne with its sub-variety Hilo; Queen, Macgragor, etc; Singapore Spanish, Red Spanish Pernamuco with a few sub-varieties; Ananas amarelo; vermelho and Monte Lirio have spineless leaves. Inventory of main pineapple cultivars is described in Table 3.

3.1 Important Varieties

A. Cayenne Group

1. Smooth Cayenne or Cayenne

It is the most popular canning variety mostly grown in Hawaii, Philippines, Australia, South Africa, Puerto Rico, Kenya, Mexico, Cuba and Formosa. The plant is stocky and

Table 3: Inventory of Main Pineapple Cultivars Included in Each Group

GROUP 1 Cayenne	GROUP 2 Spanish	GROUP 3 Queen	GROUP 4 Pernambuco	GROUP 5 Mordilonus-Perolera-Maipure

- Champaka 163
- Champaka 180
- Hilo
- 53-116
- 53-666
- F-200   (Hawaii)
	- Espanola Roja (Puerto Rico, Mexico, Cuba)
	  - 1-56 Hybrids (Puerto Rico)
	  - 1-67
	- Cabezona
		- Natal Queen (many suckers)
		- V. C. Queen
		- Ripley Queen
		- James Queen
		- Z. Queen (South Africa)
			- Pernambuco
			- Paulista
			- Beluva Amarelo
			- Perola
			- Yupi (Brazil)
				- Milagrena (Ecuador)
				- Perolera (Columbia, Peru)
				- Mariquita
				- Amarillo
				- Piampa
				- Manzana

- Cayenne (South Africa)	- Pina de Cumana - Selangor Green - Nangka - Gandol (Spineless) - Betek - Masmerah (Malaysia)	- Mac Gregor - Alexandra - Common Rough (Australia)	- Abacaxi (West Africa and many other countries)	- Tachirense - Maipure - Bumanguesa (Venezuela)
- G 25 - G 32-33 - Cayenne de Guinea - Baronne de Rothschild (Spiny leaves) (Guinea, West Africa)			- Pan de azucar - Sugar loaf (Central Amrica)	- Random (Brazil)
- Cayenne - Gaudeloupe - St Dominguo Cayenne - Champaka (West Indies)	- Castilla (Salvador)	- Mauritus - Comte de Paris - Victoria (Malaysia, Guinea, Reunion Island)	- Eleuthera (Florida) - Venezolana - Pina valera - Papelon (Venezuela)	- Monte Lirio (Central America)
- Queensland Cayenne (Australia) - Typhones (Taiwan) - 1, 2, 3, 4, 5 (Cuba) - Cayenne de Orienta (Malaysia) - Sarawak (India) - Kew				

robust, with tapering fleshy leaves up to 90 cm in length and about 6 cm in width. The upper surface of the leaves is dark green with brownish red irregular mottling above (due to anthocyanin pigment in the epidermis) and silvery gray mottling beneath with smooth straight margins, excepting near the tip and the base, where there are few small spines. The flowers are light purple with bright red bracts and their number on a straight spike ranges from 130 to 170. The fruit is cylindrical in shape and weighs between 2 and 3 kg; fruitlets or eyes are typically broad and flat. As the fruit ripens, it acquires a deep yellow to coppery yellow colour, which first appears at the base and progresses upwards to the shoulders. The flesh is firm, close textured, juicy and with a pale yellow to yellow colour at maturity. An average acid range lies between 0.5 and 1.0 per cent and total soluble solids (TSS) between 12 and 16 Brix.° Crown is normally one, is attached to fruit without a narrow neck and has loosely imbricate leaves above. Slips are on the peduncle ranging from 0 to 10 and suckers are in leaf axils ranging from 0 to 3 and reaching a length of 35–40 cm.

2. Hilo

Its cultivation is limited to Hawaii only. It is a sub variety of Smooth Cayenne, selected in Hawaii by Collins in 1960. Its plants are smaller in size, and slips are not produced. Shoots are more and develop early. Fruit shape is more cylindrical; its size is slightly smaller with larger fruitlets. Flesh colour is more yellow having higher percentage of translucence. Fruit deteriorates more rapidly after reaching full ripeness.

3. Kew

It is a famous late maturing variety of India, mostly grown for processing and canning industry. The plants are vigorous and leaves are long with straight margins. The upper surface is dark green with a superficial brownish-red mottling. And the lower surface is silvery grey or ashy grey in colour. Leaves often have a short sector of small spines at the tip and also at the base, near its attachment to the stem, where they are irregularly arranged. Fruit weighs 1.5 to 2.5 kg, and is oblong in shape, slightly tapering towards the crown. Eyes are broad and shallow, making fruits more suitable for canning. The fruit is yellow when ripe and flesh is light yellow, almost fibreless, and very juicy with 0.6–1.2 per cent acid, and its total soluble solids content varies from 12 to 16 brix. Normally fruit will have one crown but occasionally more are present. Slips arising on peduncle are 0 to 10 and number of suckers produced per plant vary from 0 to 2. This shy suckering habit is a disadvantage in its multiplication.

4. Giant Kew

This variety grown in certain regions of India is synonymous to kew excepting in size of the plant and the fruit which are larger than kew.

5. Charlotte Rothschild

This variety is under cultivation in Kerela and Goa. Otherwise, very similar to Kew in fruit characteristics and taste.

B. Queen Group

1. Queen

This is an old cultivar and grown in Australia, India and South Africa, where it is favored for taste of fresh fruit.

The plants are characterized by dwarf compact habit of growth. Foliage is bluish green. The leaves are short, stiff,

Figure 8: Var. Queen

spiny along the margin, and thickly covered with a whitish
bloom on both surfaces. The flowers are lilac. The fruit
weighs 0.9–1.3 kg. The peduncle is short; fruitlets or eyes
are small, prominent, deep set. When fully mature, the fruit
is golden yellow and internal flesh is deep golden yellow.
The flesh, although less juicy than Cayenne, is crisp (less
fibrous), transparent with a pleasant aroma and flavour.
The total soluble solids content varies from 15 to 16 brix
and acidity between 0.6 to 0.8 per cent. The slips are 0-4
and suckers are 0-3 and both are smaller in size than those
of Cayenne.

2. Ripley Queen

It is a selection from Queen and is grown in Australia.
It has pale green foliage heavily tinged with red, and fruit
is more conical in shape than queen, with a distinctly
flattened top. Fruits have pale colored skin and a richer
flavor.

3. Mauritius

It is grown in some parts of Meghalaya and Kerala in
India. Fruits are medium size and are of 2 types, deep yellow
and red. Fruits of yellow variety are oblong, fibrous, and

medium sweet compared to red type. Mauritius is exclusively grown for table purpose. Leaves are yellowish green, spiny throughout the margin. Crown also is spiny in both the types.

4. Alexandra

This variety is grown in Australia. It is a local selection made in Queens land from Ripley Queen and is similar to parent variety, but is more vigorous and produces somewhat larger fruit.

5. Mac Gregor

It is a local selection from Queensland in Australia from variety Queen. It is sturdier, broad leaved, more vigorous and produces larger and better-shaped fruits.

6. Z. Queen

This is a sub variety of Queen, reported from South Africa. It was found as a single unusual plant in a plantation of queen, and it is supposed to have originated as result of somatic mutation. It has spiny leaves of Queen but fruits are larger and have comparatively square shoulders. Texture of the flesh and color are similar to Queen. It also resembles Mac Gregory, sub variety of Queen.

7. Abacaxi

It is widely grown in Brazil for local markets. Plants are erect and leaves are 60-70 cm in length with dark green ground colour, and slightly reddish upper surface. The fruit is pyramidal in shape and weighs about 1.5 kg. Fruitlets are small and eyes are shallow. The flesh is pale yellow to almost white, tender having very small fibres, and with an abundance of juice. Juice is less acidic than Cayenne and general flavour is mild and good. The core of the fruit is small. Keeping quality of the fruit is poor unless harvested at half ripe stage. Of late, Abacaxi is considered as a separate group of pineapple. Related cultivars of this group are Pernambuco and sugar Loaf.

8. Cabezona

This is grown in Puerto Rico for trade of fresh fruits and for shipping, to New York. It is seedless, whereas Cayenne and Red Spanish produce some seeds when grown together in Puerto Rico.

C. Spanish Group

1. Red Spanish

It is extensively cultivated in West Indies, Cuba, Puerto Rico and Mexico, and is mainly used for trade of fresh fruits.

The plant and fruit size is intermediate between Cayenne and Queen. The leaves are long, about 1.2 m and spiny. Fruit is rather square in shape and weighs between 0.9 and 1.8 kg. Peduncle is long (20-25 cm) and slender and is often not able to support the fruit upright. Fruitless are few, about 80, larger than Cayenne; shell is tough and firm, and is orange red. The eyes are located deep, as in Queen group. Flesh is pale yellow, fibrous with pleasant penetrating aroma and spicy acid flavor; quite different from that of Cayenne or Queen. Core is relatively large. Crown is 20-25 cm long, with spiny recurved leaves. Slips are 2-8, and are borne very close to fruit. Suckers range from 1 to 3 per plant.

2. Singapore Spanish

It is grown in Malayasia for canning industry. This is also known as Singapore Canning and Nenasmerah. The leaves are around 50 in number. They are 100 cm long and are slightly narrower than those of Cayenne. This variety has smooth leaves with a few spiny near the tip. The fruit is cylindrical in shape, weighing about 1.6–2.3 kg. The fruitlets range between 70 and 110 and have slightly protruding surface with deep eyes. The ripe fruit is reddish orange and flesh is golden yellow, fibrous and good flavoured, contributing to the quality of the canned products. Crown is 10-30 cm. The long, frequently multiple and fascinated.

3. Masmerah

Grown mainly in Malayasia. This is a selection from Singapore Spanish. The plant resembles a typical Singapore Spanish, excepting that it is more vigrous. The leaves are generally spineless. The longest leaf is about 120 cm long and 6 cm wide at the mid point. Leaves tend to be more erect than those of Singapore Spanish. Inflorescence bears numerous purple florets, which are subtended by red bracts. Fruits are borne on a 30-40 cm long, peduncle, which is slightly thicker than that of Singapore Spanish. The fruit is cylindrical, weighs 1.5–3.0 kg, and bears 1-12 slips. As Masmerah is closely related to Singapore Spanish, the skin of the fruit is closely related to Singapore Spanish, the skin of the fruit is also thick. The flesh is translucent and intensely golden. The TSS content is about 10 brix and acidity is around 0.5 per cent. The crown is large, 50 cm or longer.

Indigenous Types Grown in India

Jaldhup and Lakhat

These are two local types, both being named after the places of their maximum production. Both fall in Queen group of fruits, being smaller than Queen. Lakhat is markedly sour in taste, whereas Jaldhup has its sweetness well blended with acidity. Fruits of Jaldhup have a characteristic alcoholic flavour of their own and can be easily distinguished from fruits of the Queen group.

Simachalam

Largely grown in Vishakhapatnam district of Andhra Pradesh.

Baruipur Local

It is largely grown in Baruipur, Sonarpur and Joynagar areas of South Bengal. The plants are moderately vigorous. Leaf margins are heavily serrated. It has heavy slip and suckering habit. Plants are very hardy and can stand adverse climates. Fruits are of small size (weighing between 1 and 2 kg), conical in shape and taper towards the crown. The crown is single and big. Rind colour of the ripe fruit is reddish yellow. Eyes are prominent, irregular and deep set. Flesh is yellow, fibrous, little stingy and sour to taste. It is neither suitable for canning nor for table purpose.

Haricharanvita

It is grown in some pockets of Siliguri sub division of Darjeeling, district and Sadar sub division of Cooch Behar district. Its plants are vigorous, leaves are long and slender, and leaf margins are heavily serrated. It has heavy slip and suckering habit. Plants are very hardy and can stand adverse climates. Fruit weight ranges from 0.75 to 2.0 kg. Fruit is conical in shape and tapers towards the crown. The rind of the ripe fruit is greenish yellow; skin very thick and eyes are prominent

3.2 Varietal Improvement

The breeding in pineapple aims at a vigorous plant,

having a short cycle and resistance to diseases (especially mealy-bug wilt); with broad, short and smooth leaves; and cylindrical fruit, well coloured, with flat eyes, on a short but strong fruit-stalk; a small ratio of leaf to fruit; with firm flesh, well coloured, not fibrous with high drymatter content, moderate acidity, high vitamin C and a narrow axis; early formation of 1 or 2 shoots and presence of 1 or 2 slips at least 2 cm below the fruit base (Etudes sur Ananas, 1977).

Cayenne is the world's major canning cultivar and many of the breeders use it because of its high proportion of good characters. Cayenne needs to be improved for its larger and more vigorous root system, more cylindrical fruits, which ripen more uniformly from base to apex, deeper yellow and crisp flesh, better quality of winter-produced fruits and resistance to wilt.

Desirable Plant and Fruit Characteristics

A good fruit is long, cylindrical and broad-shouldered, with large, flat-eyes and a small core. It should be low-set on a short fruit-stalk, bearing not more than 3 slips, all set well below the base of the fruit. The stem should be relatively short, with at least 2 suckers, originating close to the ground to ensure a stable ratoon plant. The number of suckers and the length of the fruit-stalk are influenced by the growing environment.

Although Smooth Cayenne can be regarded as an ideal type in many respects, it could be improved by selection within its variable populations. Breeding work carried out at the University of Puerto Rico by crossing Smooth Cayenne with Red Spanish resulted in 2 hybrids. They are PR 1-56 and PR 1-67. Leaves are spiny in PR 1-56 and are spineless in PR 1-67. In both average fruit weight is about

2.5 kg. PR 1-56 has white flesh and PR 1-67 has pale-yellow flesh. PR 1-56 is resistant to mealy-bug wilt and the other one is tolerant.Feng Shan branch station of Tari, Taiwan, also released 2 hybrids, Typhone No. 1 and Typhone No. 3. In both, plants are dwarf and high yielding. The fruit shape is cylindrical in Typhone No. 3.

The characteristics most commonly sought after by the canning industry and for export of fresh fruits of pineapple are listed in Table 4 (Py *et al.*, 1987)

Mass Selection Methods

The simplest method of selection is eliminating unsatisfactory plants. Defect in plant-type may be marked by lack of vigour; hence selection should be practised on a vigorous plant-crop, which matures in summer, when desirable and undesirable characters are most clearly expressed. Selection should commence about a month before the fruits reach maturity so that they can be used in assessing desirability of the plant-type.

To save time, a preliminary assessment of the area can be made to determine relative proportion of desirable and undesirable types. If the former predominates, rejected plants may be labelled, and vice-versa.

As the fruit is harvested, selected tops should be reserved for planting in the following autumn. The slips on selected plants may be left on the parent fruit-stalk until required for planting.

In-Vitro Culture

In-vitro culture of pineapple was started using apex of the crown (Mapes, 1973), slips (Sita *et al.*, 1974), axillary buds from crown (Mathews *et al.*, 1976), and more recently

Table 4: Ideal Plant and Fruit Characteristics for Canning and Export

Plant Characters		Fruit Characters	
Vegetative Growth and Development	*Required for Canning and Export of Fresh Fruits*	*Required for Canning*	*Required for Export of Fresh Fruits*
☆ Rapid growth	☆ High mean weight	☆ Broad, flat fruitlets	☆ Yellowish orange skin
☆ Semi-erect growth pattern	☆ Homogenous ripening from bottom to top of fruit	☆ Shallow blossom cup	☆ Small to medium crown
☆ Short, wide leaves	☆ Fruit well filled	☆ Flesh slightly translucent when ripe	
☆ Spineless or few spines only at the tip	☆ Firm but not fibrous, yellow flesh	☆ Small core	
☆ Less than 3 slips not closer than 2 cm to the base of the fruit	☆ Firm epidermis	☆ Large crown	
☆ Low, well-developed suckers at the base of the plant	☆ High sugar content		
☆ Short- to medium-fruit peduncle easily able to bear ripe-fruit	☆ Moderate acidity		
☆ Resistant or tolerant to main pathogens and parasites, *i.e.* mealy bugs, nematodes, *Phytophthora, Fusarium* and *PeniciHium*	☆ High ascorbic acid		
☆ Able to adapt to irregular water supply	☆ Pleasant flavour		
☆ Able to adapt to wide range of soil types			

syncarp (Wasaka *et al.*, 1978). Though plantlets are easily obtained, there is a high percentage of variation (Wasaka, 1979). Depending upon the origin of the material and the type of environment used for culture, *in-vitro* culture could be used either for obtaining variants or for accelerated propagation.

3.3 Global Pineapple Breeding Research

The overdependence of the pineapple industry on a single cultivar with a narrow genetic base has made it extremely vulnerable to the threats of pest and diseases. Development of new, resistant cultivars seemed the right strategy to redress this situation. Further, while smooth cayenne is quite good for processing, the fresh-pineapple markets in the world are more diversified and, where a choice is offered, this cultivar is not always preferred. These are primary justifications for the commencement of breeding programmes for pineapple worldwide.

In 1914, the Pineapple Growers Association of Hawaii started one of the earliest and most concerted efforts in pineapple improvement. The experimental station of this association later became the Pineapple Research Institute of Hawaii (PRI). The works of K.R. Kerns and J.L. Collins from PRI are renowned and are still often cited by present-day researchers. One of the main objectives of the PRI programmes was to develop pest and disease resistance in a "Smooth Cayenne" type variety. Varieties were developed which had better resistance to *Phytophthora*, mealybug wilt, nematodes, pink disease and internal brown spot. Some selections also had higher vitamins A and C, less acid in winter-ripened fruits, a better harvesting peak, higher yield and plant vigour and improved cannery recovery. Many varieties were selected which were better than "Smooth Cayenne" in certain individual characteristics, but eventually none could replace "Smooth Cayenne" because, after extensive evaluation, each would reveal a fatal flaw (Williams and Fleisch, 1993). Many other pineapple-growing countries have also started breeding programmes to develop high yielding varieties with specific adaptation to their own environments. Most current breeding programmes are focused on the development of cultivars for the fresh market. It is important to note that hybridization programmes attempting to replace "Smooth Cayenne" for processing have been rare since the termination of the PRI programme in Hawaii. Most hybridization programmes revolve around the "Smooth Cayenne" because of its all-round good qualities and universal acceptance. In 1926, Taiwan (then Formosa) started a breeding programme using "Smooth Cayenne" crossed with several of its local varieties, "Ohi", 'Uhi', 'Anpi' and 'Seihi'. A later hybridization programme using 'Smooth Cayenne' and 'Queen' resulted in the selection of 'Tainung 4' (Easy Peeler), the eyes (fruitlets) of which can be picked off for eating without having to peel the fruit (Fitchet, 1989). This programme continued, with the release of 'Tainung 13' and 'Tainung 16' in 1995 and 1996, respectively (Chang *et al.*, 1997).

In Malaysia, early pineapple improvement programmes focused on the selection of promising variants in clonal fields. This has resulted in the development of 'Masmerah', a higher-yielding variant of the 'Singapore Spanish' (Wee, 1974). Later hybridization programmes conducted by the Malaysian Agricultural Research and Development Institute (MARDI) developed a hybrid called 'Nanas Johor' from a cross between 'Smooth Cayenne' and 'Singapore Spanish' (Chan and Lee, 1985). In 1996, another hybrid called 'Josapine' suitable for table fruit, was developed from a cross between 'Nanas Johor' and 'Sarawak' (a variant of 'Smooth Cayenne') (Chan and Lee, 1996).

The 'Smooth Cayenne' was again featured in breeding programmes carried out in the Philippines, cote d' Ivoire, Cuba, Puerto Rico and Australia. In the Philippines, the recent breeding programme carried out by the Institute of Plant Breeding (Los Banos) was based on crosses between 'Smooth Cayenne' and 'Queen', with the objective of developing spineless 'Queen' types. One selection was micro-propagated (Villegas *et al.*, 1995). The research on pineapple in Cote d'Ivoire was started in 1978 and conducted by the fruit department (Department des Productions Fruitiers *et al.* Horticoles) of the Centre de Cooperation. Internationale en Recherche Agronomique pour le Developpement (CIRAD-FLHOR). 'Smooth Cayenne' was crossed with 'Perolera' (in fact 'Manzana') with the objectives of developing high ascorbic acid content for resistance to black-heart disorder (an internal fruit breakdown), developing resistance to phytophthora and nematodes and prevention of green ripe fruits (internal flesh colour turns yellow before the change of peel colour) (Loison-Cabot, 1987). A new hybrid called 'Scarlett' with commercial potential as a fresh fruit was selected from this hybridization programme (Coppens d'Eeckenbrugge and Marie, 2000).

In Cuba, the hybridization programme was started in 1991 using 'Serrana' ('Smooth Caynee') x 'Perolera' and three hybrids were micro-propagated for commercial fruit were micro-propagated for commercial fruit production

(Benega *et al.*, 1998b). In Puerto Rico, a natural cross between 'Smooth Cayenne' and 'Espanola Roja' resulted in a selection called PRI-67 (Ramirez *et al.*, 1972). In Australia, the breeding programme utilizes mainly "Smooth Cayenne', 'Queen' and selected PRI hybrids '73-50', '53-116' and 59-656' and the Philippine hybrid '24-116' and generally produced more hybrid progenies with commercially acceptable attributes for the fresh-fruit market.

In Brazil, the major cultivars are 'Perola' and 'Smooth Cayenne' and the main objective of the breeding programme conducted by the Empresa Brasileira de Pesquisa Agropecuaria (EMBRAPA) was to develop cultivars with resistance to fusariosis disease caused by *Fusarium subglutinans*. The resistant parents used in the crosses with 'Perola' and 'Smooth Cayenne' gave the highest proportion of hybrid progenies with desirable characters (Cabral *et al.*, 1993). Fifty-seven hybrids have been selected for further evaluation. Three 'Perolera' x 'Smooth Cayenne' and two 'primavera' x 'Smooth Cayenne' hybrids are being tested for eventual release.

Chapter 4

Propagation, Planting and Plant Density

4.1 Propagation

Propagation of pineapple like most other fruit crops is exclusively done by vegetative means. In case of hybrids, progenies evolved through seeds are also vegetatively propagated. A wide range/variety of planting material can be used to propagate pineapple plants. The most common plant parts used for propagation are suckers and slips. Crowns are also used with varying degrees of success. Plantlets arising from old plants usually carry all inherent characteristics of the parent plant. These plantlets are in the form of suckers, shoots, slips, hapas and crowns.

1. Suckers

Suckers arise and grow from buds in the axils of leaves above the ground level. They are sparsely produces. Suckers

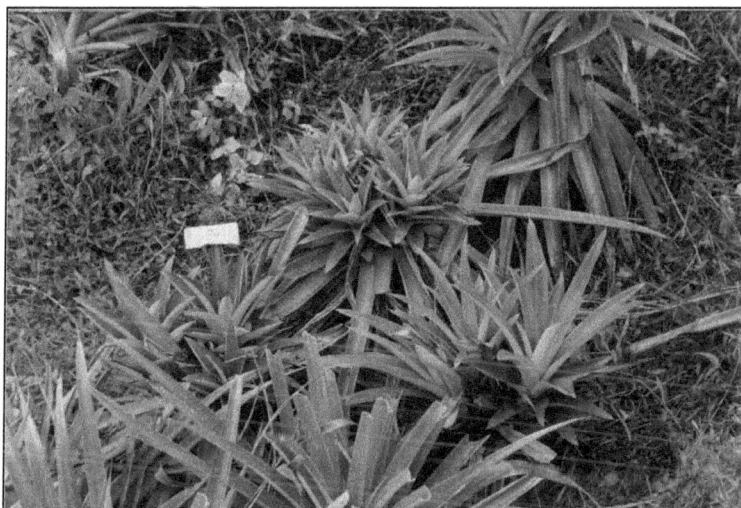

**Figure 9: Sucker Induction in Pineapple var.
Kew by Combined Spray of Paclobutrazol
50 ppm + Thiourea 250 ppm**

are grown on the plants for few to several weeks after fruit harvest. Suckers can be stored and shipped without much loss of vigour. A particular problem with sucker is that they might have under gone floral deformation before they were harvested in in shade or and storing may have induced differentiation. Because these plants do not grow before floral differentiation a small fruit is produced in a small plant.

In shy sucker bearing varieties of pine apple like 'Smooth Cyenne" for getting a second crop reduced sucker production makes it difficult for getting sufficient replanting material. Paclobutrazol 50 mg/litre and thiourea 250 mg/litre was found to produce maximum number of suckers per plant (3.86 in rainy season and 4.06 in dry season of required size and quality in least days of application (Singh and Sharma)

Singh 1997 reported that side suckers can also be increased by suppressing dominance of slips. Deslipping followed by application of 20 g nitrogen and spray of 2 per cent urea produces significant increased in desired suckers (> 500 g weight) without effecting size and quality of fruits.

2. Shoots

Shoots are leaf-branches arising from the buds in the axils of leaves above the ground level. Both of them look-like daughter plants by the side of the mother-plant. They emerge out as the mother-plant flowers, and can be seen growing fast when fruit develops. After the fruit is harvested from the mother-plant, these grow still vigorously and bear fruit in the forthcoming season. This is known as the ratoon crop.

3. Slips

These shoots are similar in origin in that they develop on the peduncle from the differentiation of lateral buds during fruits initiation. Slips grows out wards and then upward around the base of the fruit, resulting in sharp curvature at the in the base of their stem. Not all the pineapple varieties develop slips, but when they are present in large numbers, they reduce fruit weight and yield (Collin, 1960). Slips are good source of planting material but large levels should be removed before planting, as they rot in the soil and lead to rot of the stem pieces under wet conditions. The performance of slips in terms of month from planting to harvest is generally intermediate between crown and suckers because they are the intermediate in size between a small crown and large sucker. Slips arise from fruit-stalks. They are comparatively smaller than suckers, and are borne more in number per plant than suckers. They cannot grow into individual plant *in situ* and have to be transplanted in the soil. Crown grows on the top of the fruit. It is the vegetative growth at the top of the fruit, attached to the central core of the fruit. It is called crown as it adorns the apex of the fruit. Of the 3 types of planting materials, suckers are the largest material in size.

4. Discs

Fruit-stalks cut into bits known as 'discs' can also be used for propagation, but their use in field is very limited.

5. Hapas

Hapas emerge well below the fruit and have the same arrow shape as suckers initiated below on the stem.Hapas are shoots produced at the base of the peduncle. They are intermediate in size between shoots and slips.

6. Butts

Butts or stumps are the stem parts of the plant that have borne fruits. These consist of entire plants after the fruits have been harvested and from which the base of the stem, roots, leaves and peduncle have been removed. The older they are, the less suitable they will be as a plant. Butts are not generally recommended as a planting material and should only be used as the last resort.

Methods have been devised to speed up rate of asexual reproduction in pineapple for breeding work or when planting material is limited. An axillary bud is present in the axil of each leaf of a mature stem, but only 0-3 shoots are produced per plant. When some of the dormant axillary buds are cut-off with a little of the adjacent stem tissue, they can be induced to form new plants. The leaves and roots are removed and stem is cut longitudinally into 6-8 pieces. The stem is then cut transversely into triangular slices of about 2.5-cm thickness, each containing 1 or 2 buds. The pieces are fumigated to check organisms which would cause their rotting. The pieces are then planted in the sterilized shaded beds, with the inner point downwards, 5-cm apart and covered with soil to a depth of 1.2-2.5 cm. They should be watered sparingly during the first 10 days to harden off the sections. Some of the buds grow and develop roots. The plantlets are removed after 3-5 months and are transplanted into nursery beds. They will produce mature plants in 2Vi– 3 years. The buds in the axils of shoot and crown leaves can be used in a similar way.

7. Crowns

Any vegetative shoot of pineapple plant can be sectioned for propagation and the pieces have 3 or more axillary buds and some leaf material. Crowns can be inexpension, if planted fresh perform well if protected against heat and but rot with fungicides. Treatment of crown cuttings (maximum up to 8 from each crown) with copperoxychloride (0.3 per cent) + Ethrel (100ppm) is quite effective and beneficial for getting early sprouting and maximum survival of the crown cutting (Singh *et al.*, 1994).

However, crowns are not usually available because fruit is marketed with crown. Also crowns are unsuitable as planting material of the growing point has been grouped in an attempt to increase fruit weight during later stages of the fruit development.

**Figure 10: Sucker Induction in Pineapple var.
Kew by Application of Copperoxychloride (0.3 per cent)
+ Ethrel (100 ppm)**

Micropropagation

Most of the tissue culture research in pineapple relates to the use of micropropagation for the rapid multiplication of the cultivars. Traditional method of pineapple propagation usually produce up to ten plants using crown, slips and suckers from a single plant in a year. Sectioning of these components including the stem can deliver up to 100 plants. The chemical chlorfurenol can be used in commercial situations to enhance slip production as much as 30 folds. But for the sheer scale of multiplication none compares with tissue culture where according to Pannetier and Lanaud 1976, 1 million plants could theoretically be achieved in 2 years from a single axillary bud.

The first report of pineapple in vitro propagation comes from the work of Aghion and Beauchwsne (1960), and many of the earlier studies (Mapes, 1973: Mathews *et al.*, 1976: Dnew, 1980) concentrated on culture establishment and the ability to take plants successfully through the micropropagation process. Subsequent studies have concentrated on methods to optimize multiplication and today pineapple micropropagation is being used commercially in the pineapple industry to rapidly multiply important cultivars (Smith and Drew 1990). In practice micropropagation is used for the establishment of multiplication blocks, which then provide conventional planting material for larger production blocks. This is because micropropagated plants are more expensive and there are grower concerns about genetic off type (somaclonal variants). The initial limited use of micropropagation in the multiplication of new cultivars allows screening of variants on the fruit characteristics before conventional methods of multiplication are used. The most common explant used to initial cultures is the axillary bud dissected from crown leaves. Fitchet 1990 also suggested that the crowns of some cultivars, such as smooth cyenne should first be desiccated for a short period to break bud dormancy. Clean buds are most commonly grown on a Murashige and Skoog (1962) (MS) solid medium supplemented with a cytokinin, usually benzyladenine (BA) between 2 and 5 mg^{-1}. Using these concentrations with a cultivar such as " Smooth Cyenne" 10-15 plants per month can be produced. Multiplication can be enhanced 2-3 fold by the use of agitated liquid media (Mathews and Rangan 1979; Moore *et al.*, 1992 and further refinement have been made by the use of temporary immersion systems (Firoozababy *et al.*, 1995:Escalona *et al.*, 1998). A novel propagation method was developed by Kiss *et al.* (1995); this involved the induction of etiolated shoots, with subsequent multiplication of shoots along the nodal systems when placed horizontally on the culture medium.

Once shoots have been produced and multiplied they are usually transferred to an MS solid medium containing an auxin, such as indole butyric acid (IBA), at between 0.5 and 2.0 mg^{-1}, or to a harmone free medium for root development. The use of *Azotobacter* or endomycorrhzal fungi has also been suggested as a way of improving the growth of micropropagated plants (Gonzalez *et al.*, 1996; Guillemin *et al.*, 1996; Matos *et al.*, 1996).

Plants can also be regenerated from cell or callus cultures. When the cytokinin is supplemented with an auxin, such as Napthalene acetic acid (NAA), the development of lumpy tissue with protocorm–like bodies, or callus tissue can result (Mathews and Rangan, 1979, 1981;Wakasa 1989; Devi *et al.*, 1997). Proliferation of these type of tissues often results in cultures with an enhanced regeneration capacity via either organogenesis (Fitchet, 1990) or embryogenesis (Daquinta *et al.*, 1996; Cisneros *et*

Figure 11: Micropropagation of Pineapple

Table 5: Review of Methods Used for the Micropropagation of Pineapple

Stage I	Stage II	Stage III	Stage IV	Reference
Explant: Axillary and terminal buds from crown Medium: Nitsch with 0.1 mg l⁻¹ BA and C. 1 mg l⁻¹ NAA	Medium: MS with 2.0 mg l⁻¹ KIN, 2.0 mg l⁻¹ IBA and 1.8 mg l⁻¹ NAA 8 plantlets per month	Medium: MS with 0.1 mg l⁻¹ NAA and 0.4 mg l⁻¹ IBA	'Soil'	Mathews et al. (1976)
Explant: Axillary buds from slips and suckers Medium: MS, hormone-free	Medium: MS with 2.3 mg l⁻¹ KIN 50 plantlets per month (callus involved)	Not Stated	'Gro-pots'	Drew (1980)
Explant: Axillary bud from crown Medium: MS with 25 per cent coconut water	Medium: MS with between 0.5 and 1.0 mg l⁻¹ BA 3 plantlets per month	Medium: ½ MS, hormone-free	Not stated	Zepeda and Sagawa (1981)
Explant: Axillary bud from crown Medium: MS with 2 mg l⁻¹ BA and 2 mg l⁻¹ NAA	Medium: As for stage I, but liquid (shaken) cultures 17 plantlets per month for 'Smooth Cayenne'; 76 plantlets per month for 'Perolera'	Not roots	'Commercial soil mix' plantlets > 2.5cm	DeWald et al. (1988) Moore et al. (1992)

Contd...

Table 5–Contd...

	Stage I	Stage II	Stage III	Stage IV	Reference
	Explant: Axillary bud from crown Medium: MS with 0.5 mg l^{-1} BA and 0.2 mg l^{-1} IAA	Medium: MS with 0.5 mg l^{-1} BA 'Plantlets halved or quartered during subculture' 10 plantlets per month	Medium: MS, hormone-free	'Peat/perlite mix' Plantlets 2-3 cm	Cote et al. (1991)
	Explant: Axillary bud from crown Medium: MT with 2.0 mg l^{-1} KIN, IBA and NAA	Medium: MT with 2.0 mg l^{-1} KIN and NAA 14 plantlets per month	Medium: MT with 1.0 mg l^{-1} NAA and 500 mg l^{-1} malt extract	'Peat/perlite/sand'	Fitchet (1990) Fitchet-Purnell (1993)
	Explant: Axillary bud from stem Medium: MS with 2.3 mg l^{-1} BA and 0.6 mg l^{-1} NAA	Medium: As for stage I 8 plantlets per month	Medium: MS with 0.3 mg l^{-1} IAA	Not stated	Osei-Kofi and Adachi (1993)
	Explant: In vitro plantlets Medium: MS with 2.0 mg l^{-1} NAA in the dark	Medium: N6 with 5 mg l^{-1} KIN and 4.5 mg l^{-1} BA 'Etiolated shoots placed horizontally on medium' 60 plantlets per month	Medium: MS, hormone-free	'Soil' Plantlets 8 cm	Kiss et al. (1995)

Contd...

Table 5—Contd...

	Stage I		Stage II		Stage III	Stage IV	Reference
Explant	Apical bud from crown	Medium	As for stage I, except 2.0 mg l⁻¹ BA	Medium	MS with 2.0 mg l⁻¹ IBA	Not stated	Devi *et al.* (1997)
Medium	MS with 1.0 mg l⁻¹ BA and 0.1 mg l⁻¹ NAA		'Regeneration from proliferating callus at base of plantlets' 56 plantlets per month				

BA: Nenzyladenine; IAA: Indole acetic acid; IBA: Indole butyric acid; KIN: Kinetin; NAA: Naphthalene acetic acid; MS: Murashige and Skoog (1962); MT: Murashige and Tucker (1969); N6, Chu (1978); Nitsch (1951).

al., 1998). Regeneration of plants from cell and callus transformation of pineapples. However, these procedures must be viewed very catiously because of perceived problems with somaclonal variation. As Scowcroft (1984) pointed out, as regeneration proceeds from more organized structures (axillary buds) to unorganized tissues (callus), the propensity of the cells to undergo genetic changes increases.

While assessing the pineapple plants developed from micro propagation instead of conventional suckering. Singh and Mondal (2000) reported that micro propagated plant lets took more time to flowering and harvesting but they flowered well. The quality of the micro propagated plants was comparable with those from suckers.

4.2 Choice of Planting Material

Performance of plants as characterized by vigour, growth rate, time taken for bearing and fruit size and quality varies with the type of material used for planting in pineapple. Besides the type, size of the planting material also results in the variation in the performance of the subsequent plants. With different types and sizes of planting materials, a number of management difficulties crop up on account of the poor rate of plant establishment, uneven growth of plants, and uneven flowering and harvesting, stretched over a long time. Besides, uniform cultural operations cannot be taken up, and ultimately plant-wise operations have to be followed, which margins the efficiency of the labour and other inputs, resulting in increased cost of production. In a mixed planting, a few plants are in flowering while the others are ready for harvest. This state of plant growth poses problems for getting good uniform ratoon crops also. Therefore, it is advisable to use uniform-size material of a monotype for uniform growth of plants. Hence, it is imperative to know, what is the right type and size of planting material to be selected for commercial planting. Studies carried out at Basti (Teaotia and Pandey, 1966) and Bangalore (Chadha *et al.*, 1974) have indicated the superiority of slips over suckers; which in turn are found better than crowns. Both in suckers and slips, larger planting material resulted in vigorous plants. Neither too big nor too small slips or suckers resulted in more flowering, and the large grade material resulted in the more staggered flowering period. Smaller was the slips or suckers, earlier was the fruit maturity. Slips yielded quality fruits and yields were at a par with the highest yield out with the largest suckers. Of the 3 types tested, crowns resulted in least vigorous plants. Their harvesting took more time. Besides, survival percentage was less and yield was poor. But fruits were of good quality. In the overall analysis, slips weighing from 450 g were reported to be the best planting material.

Among the types and sizes of propagules tried, slips and sin weighing around 350 g and 450 g, were found the best in terms of and quality for Kew pineapple under Coorg (Karnataka) cond (Singh *et al.*, 1978). In Jorhat (Assam), suckers weighing 501-750; slips weighing 301-400 g were found ideal planting material (AICPFI, 1982), while suckers weighing 501-1,000 g were best for pla in Thrissur (Kerala) (Varkey *et al.*, 1984).

It can be concluded that in the event of non-availability of suckers weighing around 500 g are ideal, and in the event of availability of sufficient number of suckers, due to poor suck habit of the varieties like Kew, slips weighing around 350 g are best.

However, shortage of propagules in pineapple is felt universally. Mass multiplication of propagation material is vital to bring fresr under cultivation. This is possible only when a number of plantlets can be obtained from a single mother-plant, unlike a few suckers or slips. Possibility of using leaf-cuttings from crowns of cvs Kew and smooth Cayenne for multiplication of planting material has been shown. Fifteen leaf-cuttings are made from each crown. However, these cut will take even more time than crowns for flowering and thus are recommended when sufficient planting material is not available (Dass *et al.*, 1976; Venter 1988)

Sectioning

Any vegetative shoot of the pineapple plant can be sectioned for propagation if the pieces have 2 or more axillary buds and some leaf material. For large stems, the leaves are cut off, the stem are quartered length wise and each piece is then cut in to sections 3-5 cm in length. A small amount of leaf material is usually left attached to each piece. The sections are cured by drying, or treated with fungicide, or both, and planted in a well prepared nursery. The soil in the nursey should be fertile at the time sections are planted or new shoots can be fertilized by foliar feeding with a dilute nutrient solution. Crowns may be similarly micresectioned, producing four or more sections per crown. When such sections are produced, they are best planted in the fertile media and grown in the green house. Small crown sections should be started in a well drained medium, such as course sand, and must be kept moist. When using a medium of low native fertility, foliar feeding should commence soon after the new shoots emerge. All materials produced by sections must be grown to an adequate size for field planting and hardened if they have been grown under shade or green house conditions. Plantlets may developed rapidly in the field if they have a well developed root system at the time of planting.

Gouging

Gouging is the mechanical removal of the shoot apex of growing plants, usually with a tool specially designed for the purpose (Heenkendra, 1993). The young leaves are pulled out prior to gouging. Gouging destroys the plants apical dominance, thus releasing axillary buds from their induced dormancy. These buds form shoots, which, when large enough are harvested and planted in a nursery or a field.

Plants as small as 100gm have been gouged but such small plants would normally be grouped in a green house with the fine tools. When grouping is done in the field, plants should be larger so that gouging can be done by workers from a standing position. The gouge should be shallow preferably not more than about 1.0 cm below the apical bud. Treatment of the gouged plants is necessary to avoid rot. Shoots produced after gouging are allowed to grow to the desired size and their mass at the time of removal depending upon the planting conditions. Very small plants less than 50 gm or so would normally be planted in a green house under shade. Plants weighing between 50 and 100 gm would be planted in a propagation bed in the field, while plants weighing 200 gm or more are suitable for suitable commercial planting. A well maintained mother plant nursery can be productive for 8-12 months after which it can either be rejuvenated or nocked down and replanted. Rejuvenation involves allowing atleast one low attached shoot to grow until it is of gouging weight. When expanding clones, great care must be exercised to keep the area free of pests and diseases, as the material is the foundation stock for the form.

4.3 Planting

Suckers and slips are usually preferred for planting since they flower comparatively earlier than crown. Propagation by crown is very limited, and use of stumps or discs for planting is very rare in India.

Time of Planting

The time of planting dictates the season at which the first plant-crop is obtained. Planting time is very important in view of the natural flowering period, which differs from region to region. By the time of flowering, if plant does not attain optimum physiological maturity, either it will escape flowering and flower in the next season, or if flowering is induced in the same season, the plant will bear small fruits. Hence the ideal time of planting would be 12-15 months before the peak flowering season under natural conditions, which varies from December to March in different regions. Time of planting also varies from place to place, depending upon the onset of the monsoon and intensity of precipitation. Pineapple is mainly planted just at the onset or at the end of the monsoon in order to avoid heavy precipitation in the pre-establishment period of the plants. For western India, Gandhi (1949) recommended planting during September-November. In Assam, planting is done during August-

October, and in Kerala during April-May. The best time for planting in Chethalli (Coorg) was found to be April-June (Chadha, 1977). A late planting in September delays crop at least by 7-9 months. The peak flowering under this condition was observed during January-March. Best planting time in north Bengal district is October and November, and in the other districts it is in June-July.

Pre-treatment of Planting Material

All the planting materials are subjected to variety of treatments prior to planting. These include curing, bundling, transportation from the field or growing area, grading storage, dipping, transport to the planting field, spreading and finally planting. Planting material size and uniformity are particularly important. Within any category of plantinfg material, large pieces will usually grow to forcing weight faster than smaller pieces generated in the same field. For this reason grading of planting material by size is critical to provide uniform plants at forcing and efficiencies at time of harvest. Crowns, slips, hapas, suckers and propagules are very tolerant of storage and may be stored for months between harvest and planting. However stored plants continue to grow, stem diameter decrease, and storage reserves are consumed, so prolonged storage slows initial growth and increases variability (Py *et al.*, 1987). Therefore fresh planting is almost always superior to store material because it grows faster. However very fresh material should be treated with a fungicide to avoid black rot caused by *Chalara paradoxa*. Dipping of planting material is done to protect against rots and pests.

Storage of planting material should be under conditions that allow for good air movement to avoid rot. Storage in piles or other conditions that block exposure to light should be kept to a minimum, as stem etiolation will occur if material is deprived of light for more than 2 weeks. Spreading of planting material in the fieldshould be done to facilitate the work of planters.

Before planting, suckers are sorted out into larger, medium small to avoid competition between plants of different sizes. Unit of planting material must be ascertained in a plot for the ease of cut out a particular cultural operation at a time. The plant material by removing scaly leaves from about 2-5 cm of the stem base to root initials and by trimming lower end of the stem with a keeping it exposed for 4-5 days before planting. After removing leaves, planting material should be treated with monocrotophos and carbendazim (0.1 per cent) solution to protect against mealy bugs and heart rot, respectively. Stripping of lower leaves facilitates initial rooti establishment of plants.

For getting pineapple production round the year sequential planting of different size suckers (350 and 500g) from May to December produced round the year supply of pineapple. However, planting of large size of suckers (500 g from June to Oct. showed vigorous vegetative growth, increased sucker production and maximum percentage of flowering both in main and ratoon crop (Singh *et al.*, 1998: Singh and Bandyopadhyaya 1999).

Planting System

The planting system varies depending upon the topography and rainfall. There are 4 planting systems in vogue, *viz.* flat-bed furrow planting, trench planting and contour planting (Chadha Before planting, land is ploughed thoroughly, followed by harrow give fine tilth to soil.

Flat-bed Planting

It is commonly followed in West Bengal, in plains, where there is no problem of soil erosion. Planting is dibbling in pits 15-20-cm deep. Provision is made to drain-off water in high rainfall areas and to irrigate in scanty rainfall area bed planting is also practised in irrigated areas in Bombay pro-western India (Gandhi, 1949).

Furrow Planting

This is followed in coastal plains of Kerala natural precipitation is moderate, fairly distributed, but not in and crop is grown without irrigation. Planting in this system is < 30-45 cm deep furrows alternating with 1.2-1.5-m broad ridges.

Trench Planting

Trench planting envisages at sideward anchor plants. Dhareswar (1950) recommended growing of pineapple 1 cm deep trenches in view of the soil moisture conservation. Ge about 30-45-cm deep trenches are opened at a distance of 1.0-1.5 kinds of trench planting such as single-row and double-row are of treble-row and four-row trench plantings are also seen occasionally single-row trench system is common in hill-side plantations of Assam and Tripura, where soil slope is gentle and soil erosion problem because of the moderately heavy rains. Double-row very widely followed and popular in Karnataka, Kerala, Assam,Tripura, West Bengal and Goa and wherever plantations are in plains, whether crop is grown with or without irrigation. The field is laid out into trenches alternating with mounds. Trenches are made always across the slope. In each trench, 2 shallow furrows about 10-15-cm deep and 15-cm inside from the edge of the trench are opened and suckers or slips are planted in these furrows. The plants in two rows within the trench are so arranged that they are not exactly opposite to each other.

The advantage of tri-row trench and four-row trench over double-row trench is that more number of plants can be accommodated per unit area. Experiments at the Indian Institute of Horticultural Research have revealed that the double-row trench system is better than three-row and four-row trench systems in view of the convenience in farm operations (Dass *et al.*, 1976). The depth of the trench should be about 22.5-30.0 cm (Reddy and Prakash, 1982; Radha *et al.*, 1990), and this is important for plant-crop than ratoon crops.

Contour Planting

This system is commonly seen in hill-side plantations of Assam and Tripura, where rainfall is high and the soil is subjected to erosion. Planting is done in terraces cut by digging contour trenches, which catch run-off water along with silt and serve as drainage channels. Spacing between 2 contour parallel lines of plants and width of the terrace depend on the degree of slope. Pineapple being a shallow feeder should not be planted very deep. About 10-12 cm of the basal portion of the sucker/slip should be below the soil while planting. Deep planting should also be avoided to prevent soil entering into the heart of the plant, which can kill the plant. While planting, sucker/slip should be set firm by pushing soil around and pressing it hard.

In most of the areas neat planting of pineapple is practised. Dhareswar (1950) and Naik (1963) recommended growing pineapple under partial shade of coconut and young mango trees in plains of south India, where sun is severe and rainfall is scanty.

4.4 Plant Density

Plant spacing or density of pineapple depends on the growth of the plant and system of planting. Adoption of low planting densities has been the major constraint in India, contributing towards[1] high cost of production per tonne of pineapple. One of the ways to reduce cost of production is to increase yield per unit area by following high-density planting.

For Assam, Chowdhury (1947) recommended a spacing of 60 cm within a row, 90 cm in between the rows and 120 cm in between the trenches, accommodating 15,000 plants/ ha. In western India, the practice is to accommodate about 10,000 plants/ha under flat-bed system at a distance of 90 cm x 90 cm (Gandhi, 1949). In Kerala, plants are spaced at 45-60 cm within a furrow or trench opened, 1.2-1.5-m apart under the furrow planting or single-row trench planting, accommodating 14,000 plants/ha. In double-row trench, plant population varies from 15,000 to 20,000/ha (Chadha, 1977). Unlike India, planting density followed in other pineapple-growing countries is much higher.

In Hawaii, plant density followed is 40,000-45,000 plants/ha. In the Philippines, plant density of 62,000 plants/ha is commonly followed (IIFT, 1968). Wee (1969) reported as high as 71,758 plants/ha to be optimum for Singapore Spanish variety under Malaysian conditions. Planting more than 45,000 suckers/ha was found to reduce fruit size (Collins, 1960). While total yield of pineapple increases with increased plant population, fruit size reduces with closer spacings. An ideal fruit size for canning is 1.8-0.2 kg, without crown.

According to Waithaka and Puri (1971), the yield of grade I fruits decreased with increasing population density beyond 60,000 plants/ha, and yield of grade II fruits increased steadily. Trials conducted at the Indian Institute of Horticultural Research, Bangalore, have revealed that fruit length, breadth and canning ratio did not vary significantly with varying plant densities ranging from 43,560 to 63,758 plants/ha (Fig 8; Chadha *et al.*, 1973).

Spacing followed for 63,758 plants/ha was 22.5 cm from plant to plant, 60 cm from row to row and 75 cm between beds. Fruit size, quality and canning ratio were not adversely affected by this planting density. The possibility of reducing spacing further was explored at different centres under different agroclimatic regions of the country under the All-India Co-ordinated Fruit Improvement Project. Population densities ranging from 49,300 to 111,000 plants/ha were compared. Dass *et al.* (1978) recommended a spacing of 25 cm × 60 cm × 90 cm between plants, rows and beds for Kew pineapple under two-row system, resulting in a plant density of 53,300/ha in Bangalore, keeping in view of the ease of with increasing plant densities up to 111,000/ha. A plant density from 53,300 (25cm x 60 cm x 90 cm) to 59,200 (25 cm x 60 cm x 75 cm) was also found ideal, both in irrigated and rainfed areas of Coorg, Karnataka (Dass *et al.*, 1978).

Decrease in fruit weight was quite evident when plant-to-plant spacing was reduced to 20 cm and row-to-row to 40 cm irrespective of the spacing between beds. Plant densities of 53,300 and 59,200 plants/ha were found ideal for Kerala as against traditional planting density of 15,000 plants/ha with a spacing of 45 cm x 60 cm x 180 cm in rows, between rows and between trenches. Highest cost: benefit ratio of 1:4.2 was observed in planting densities of 53,300 and 59,200 plants/ha at Thrissur (Kerala) (Annual Report, AICFIP Cell 1,1978), but the density at 59,200 plants/ha was found inconvenient for carrying out cultural operations as the inter-trench spacing is reduced to 75 cm only as compared to 90 cm in 53,300 plants/ha. A plant density of 53,300 plants/ ha was recommended in Jorhat (Assam) also (AICFIP, 1978). The average fruit size was reported to be reduced with a plant population of 111,000/ ha in Bangalore, Coorg, Thrissur and Jorhat. However, fruit size did not differ much with varying plant populations ranging from 49,300 to 60,000/ha. In high rainfall and fertile areas of West Bengal, planting densities varying from 43,500 to 49,300 were recommended, and for the rest of the areas, 25 cm x 60 cm x 90 cm spacing was recommended, accommodating 53,300 plants/ha. However, Mukherjee *et al.* (1982) recommended 63,758 plants/ha spaced at 22.5 cm x 45 cm x 90 cm for Queen pineapple in Baruipur (Calcutta) as well as for plains of south Bengal for high yields of good quality fruits and for more suckers and slips production per unit area. Trials conducted at Pineapple Research Station, Nayabunglow, Meghalaya, have shown that 43,500 plants/ha were ideal (Khongwir and Das, 1978) (Table 6).

At Hessaraghatta, rate of ratooning was compared in traditional and high-density planting of 55,555 plant/ha. In this, 2 successive ratoon crops harvested at 12 monthly interval amounted to 50.7 and 53.8 per cent of the plant crop yield (Chadha, 1977). The yield of the first and second ratoon crops is to an extent of 50-60 per cent and 40 per cent of the plant-crop in Hawaii and Australia.

Table 6: Spacing for Different Plant Populations per Hectare

Plant Population per Hectare	Distance (cm)		
	Plant-to-plant Within a Row	Row-to-Row	Trench-to-Trench
43,500	30	60	90
53,300	25	60	90
63,700	22.5	60	75
		or	or
22.5	45	90	

High-density planting, besides increasing the yield, is associated with other advantages like less weed infestation, protection to fruits from sun-burn and increased production of suckers and slips per unit area, and non-lodging of plants. Close planting also saves on the cost of providing shade to fruits, as it provides natural shade through upright orientation of the apical leaves, and eventually results in uniformly-coloured lustrous fruits. In Assam, about 20-25 per cent of fruits develop sun-burn in the absence of adequate shade. High-density planting also results in overlapping of basal leaves forming a sort of natural covering over the soil; preventing evaporational losses and thereby resulting in moisture conservation. Under dense planting a micro-climate with high humidity will be created around the plant which is congenial for growth and fruiting. It is also possible to take at least 2 ratoon crops under high-density planting. High-density planting is recommended to accommodate as many number of plants as possible while ensuring sufficient space to carry-out cultural operations.

Soil moisture and fertility influences plant growth and indirectly determines spacing required per plant and eventually planting density. In overall analysis, a plant-to-plant spacing of 22.5 cm and a row-to-row spacing of 45 cm appear ideal. Where pineapple plants grow luxuriantly with long leaves, a wider spacing of 90 cm between the trenches seems to be required, but in places where growth of the leaf is moderate, a trench-to-trench spacing of 75 cm appears adequate. However, two row-trench system of planting has been found to be the best universally.

A plant density of 63,000 plants/ha (22.5 cm × 60 cm × 75 cm) has been found ideal in sub-tropical and mild humid areas of Bangalore. In hot and humid areas of Kerala, Karnataka and West Bengal, a plant density of 53,000 plants/ha (25 cm × 60 cm × 90 cm) performs well. In rainfed, highly fertile and hilly areas, like north-eastern states, a plant density of 43,000 plants/ha (30 cm × 60 cm × 90 cm) has been recommended. The yields recorded under high-density planting are in the range of 70-105 tonnes/ha. The increase in yield per unit area is in the order of 45-85 tonnes/ha. In nutshell, adoption of high-density planting does not hamper fruit size, quality and canning recovery.

Chapter 5

Nutrition

5.1 Plant Nutritional Status

The nutritional status of the pineapple plant has a large influence on plant growth and, consequently, on yield and fruit quality. For pineapple, plant indicators that reflect plant nutritional status have been identified and, in conjunction with soil analysis, can be used to manage fertilization of the pineapple crop. To sustain growth and obtain good yields, it is important to provide adequate supplies of all nutrients in proper balance. Balanced nutrition based on the principles of best management practices ensures that excess nutrients of one type do not induce deficiencies of others or, in the case of N and P, lead to environmental degradation.

The plant indicators of nutritional status include visual deficiency symptoms and the critical nutrient levels in appropriate reference-plant tissues. This tissue in pineapple is the 'D' leaf, which in most cases is the tallest leaf on the plant. In an actively growing plant, the A, B and C leaves are older than 'D' leaf, and 'E' leaves are comparatively younger (Figure 12). The 'D' leaf is usually the longest leaf and is easily selected by gathering upper leaves in both hands. There are generally 2 and 3 leaves of similar length.

The 'D' leaf is removed by holding leaf in hand and giving leaf a sharp twist to one side and back to other side. The leaf should break out clearly with a sharp cracking sound. If it is difficult to remove then the leaf selected is probably too old and fibrous at the base.

Other ways of identifying 'D' leaf are by the shape of the leaf base, the amount of the basal white tissue and the absence of fibres (Figure 13).

The D-leaf is used because it is the only leaf that can be consistently identified and, as the youngest most

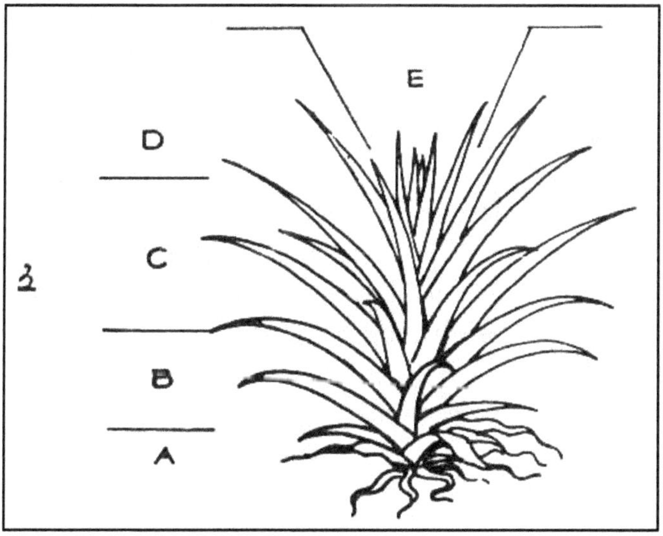

Figure 12: Leaf Classification

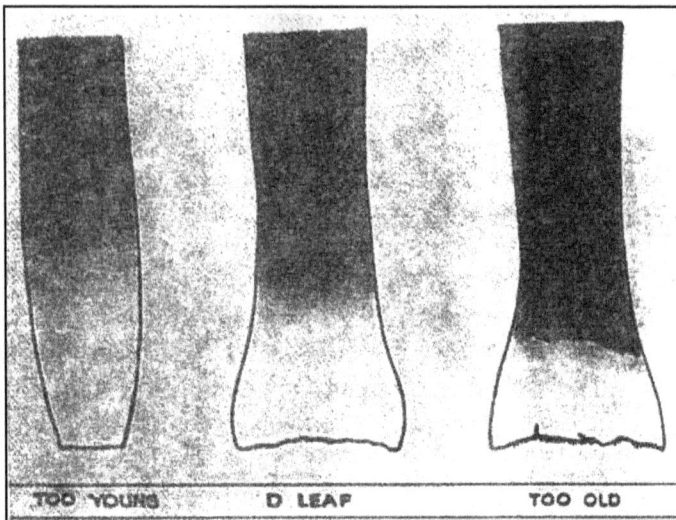

Figure 13: Classification of Pineapple Leaves

physiologically mature leaf, it reflects current plant nutrient status with acceptable accuracy. Sideris and Krauss (1936) categorized leaves as: 'A', present on the propagule at planting and do not elongate after planting; 'B', present on the propagule at planting and elongate after planting; 'C, leaves that develop after planting and are younger than 'B' leaves but older than 'D' leaves; 'D', a whorl of three leaves, including the tallest on the plant; 'E', a whorl of three leaves younger than the 'D' leaves; and 'F', a whorl of three leaves younger than the 'E' leaves. As the plant grows, leaf classification continuously changes, so that 'F' leaves become 'E' leaves, 'E' leaves become 'D' leaves, and so on. A knowledge of the visual symptoms associated with the deficiency of a specific nutrient can help with early detection of nutritional problems in the field. Nutrient levels in leaf tissue provide information about the quantity of nutrients actually absorbed by the plant. However, nutrient deficiencies can have multiple origins, which need precise identification, and several nutrient-deficiency symptoms are not diagnostic.

5.2 Diagnosis System

Of the 16 nutrient elements essential for plant growth, nitrogen (N), phosphorus (P), potassium (K), calcium (Ca), magnesium (Mg), sulphur (S), iron (Fe), zinc (Zn), copper (Cu) and boron (B) have a distinct role in metabolism of pineapple, in the production of dry matter and as a constituent of different enzymes. The morphological and physiological peculiarities of pineapple-plant could deviate the normal role of these nutrients in plants.

The nutritional status of a pineapple plant depends on many factors, including the nutritional status of the propagule, soil nutrient status, the physical and mineral characteristics of the soil, soil water status, root-system development and functionality and numerous physical and biological factors that can influence the efficiency of the root system in extracting soil nutrients. Pineapple readily absorbs all nutrient elements through the leaves (Py *et al.*, 1987), but N, P, K, Mg, Fe, Cu, Zn and B are the ones most commonly applied in solution foliarly (Swete Kelly, 1993). Calcium is normally not applied foliarly because most salts of Ca are relatively insoluble or would render other nutrients in deficient. Where foliar absorption replaces soil uptake, nutrient utilization is still dependent on the physical and biological factors that affect the extraction and utilization of plant nutrients in the soil.

The concept of a pineapple crop log was established in Hawaii in the 1940s (Sanford, 1962), and comprehensive knowledge of nutrient management was developed in many countries in the 1960s and 1970s. The crop log includes soil and plant indices as well as those biological and physical factors of the pineapple crop environment likely to retard growth and provide growers with the information required to adjust fertilizer applications to fit the requirements of the crop. However, no integrated diagnosis database capable of synthesizing this information is available.

To ensure that growth is optimum, it is essential to identify all the factors that directly or indirectly affect the uptake, translocation and utilization of nutrients. Soil indices always include the levels of P, K, Ca and Mg and pH, but may also include soluble Al and Mn and salinity. Where nitrate in fruits is a concern, soil N may also be analysed (Swete Kelly, 1993). When properly developed and calibrated, soil analyses provide an estimate of the quantity of fertilizer to be applied at planting and during early growth, indicate the levels of nutrient element reserves in the soil and indicate possible toxicities that could restrict plant growth. Opinions about soil nitrogen analyses vary.

Pineapple leaves can change quickly and so were not found to be a reliable measure of long-term plant nitrogen status. Tissue element norms for P, K, Ca and Mg for 'Smooth Cayenne' were developed for the middle one-third of the white basal tissue of the 'D' leaf in Hawaii and Australia, because the results can be expressed on a fresh mass basis, thus simplifying calculation of the final results. The fresh mass of basal tissue can be used because changes in the dry-matter content of this tissue with changing plant water status are small, typically only 1.0-1.5 per cent (Sanford, 1962).

Tissue element norms have also been developed for the entire 'D' leaf (Py *et al.*, 1957) because it was believed that nutrient element levels in leaf basal tissue were more indicative of movement within the plant and were too sensitive to daily rhythms and to variations in the soil solution. Because whole-leaf water content can vary significantly, results must be calculated on a dry-mass basis. Availability of nutrients to plants to some extent differ with cultivars. Optimum levels of N, and K in 'Red Spanish' pineapple appear to be slightly higher than those for 'Smooth Cayenne' (Samuels and Gandia-Diaz, 1960). The nutrient content of any tissue depends on its physiological stage. However, because approximately 80 per cent of a typical vegetative pineapple-plant fresh or dry mass is leaves. Young, actively growing leaves (leaves younger than the 'D' leaf) generally have higher nutrient levels than do older leaves that are fully grown (Sideris *et al.*, 1943; Sideris and Young, 1945, 1946). The nutrient content of any given leaf increases from the base (youngest tissue) to the tip (older tissue) (Sideris *et al.*, 1943; Sideris and Young, 1945, 1946). Leaf dry-matter contents of the basal portions of 'C' leaves are higher than in younger leaves and the dry-matter content of all leaves increases from the base to the tip, while that of the stem decreases from the base to the tip.

The total amount of nutrients in the plant increases with age but the concentration in any given tissue can increase, decrease or remain unchanged, depending on the nutrient and the tissue. Assuming no other limitations, nutrient concentration in the plant tends to reflect the supply available, whether provided by foliar feeding or by the soil, especially where the supply is limiting. When fertilizer applications are made, 'D'-leaf nutrient concentration may temporarily increase in response to those applications, even though the long-term trend of nutrient concentration in the 'D'-leaf basal tissue is downward with increasing time after planting. Seasonal changes in tissue nutrient concentrations occur (Sanford, 1962), but they may result from changes in the plant growth rate rather than from changes in nutrient availability to the plant Visual deficiency symptoms of each nutrient were reproduced for pineapple in the early 1960s by growing plants in single-nutrient cultures in solution or sand, with all other nutrients but the one being studied kept at optimum levels (Sideris and Young, 1951, 1956; Cibes and Samuels, 1961; Tisseau and Tisseau, 1963), although 'Singapore Spanish' (Kanapathy, 1959) and 'Red Spanish' (Cibes and Samuels, 1958) were also studied. These visual symptoms and the leaf concentrations of the nutrients associated with them are of use in diagnosing nutritional problems in the field. However, multiple deficiencies may occur in the field, which can make diagnosis difficult without confirming plant-tissue analyses, as well as an analysis of the plant and its soil and aerial environment.

5.3 Major Nutrients

5.3.1 Nitrogen

Nitrogen is an important constituent of proteins and their precursors, which forms about 7 per cent of dry matter of the pineapple plant. Nitrogen is required by pineapple in greater amounts than any other nutrient except potassium. Providing adequate supplies of N to rapidly growing plants is essential to maintain high rates of growth, ensuring vital processes of the plant and produce good yields. Application of nitrogen promotes formation of leaf materials and increases weight of the stem but does not increase root formationin proportion to the weight of the whole plant (Sideri and Young 1946). In addition to increase in the weight of the fruit, its diameter also increases (Py *et al.*, 1956, 1957a); probably through increased plant growth as the reserves are translocated from stem and other vegetative parts into the developing fruit. Besides flower formation depends on the ratio of the nitrate to carbohydrate in the plant, and it retards with increased application of nitrogen. Level of the nitrogen in D-leaf must remain at more than 1 per cent of the total dry matter (Martin-Prevel, 1959a).

Plant indices for nitrogen include leaf colour and nitrate-nitrogen in leaf basal white tissue, total nitrogen in green tissue and leaf chlorophyll content. Soil indices for N, if used, only provide an indication of the N required for early growth of plants. Many tropical soils contain small amounts of nitrogen, so meeting crop N requirements is an important and challenging task.Leaf colour is an important diagnostic index for nitrogen and is the most important index defined in the pineapple crop log of Nightingale (1942a). When N is defcient, leaves are yellowish green to yellow. However, symptoms in the field are different from those normally found in solution culture. Normally a well-exposed nitrogen-deficient plant will have yellow older leaves because the nitrogen in those plants is translocated to the younger leaves. In the case of nitrogen-deficient plants grown in the field, the older leaves remain green despite the removal of nitrogen from them because of mutual shading of lower and older leaves by adjacent plants. In the field, it is the younger leaves that are yellow. Thus, leaf colour is an integrated index that estimates the nitrogen requirement of pineapple on the basis of its relative carbohydrate status (Nightingale, 1942a).

Table 7: Characteristics of the Four Ranges of Pineapple Leaf Colour (Nightingale, 1942a; Sanford, 1962).

Designation	Colour	Cell Starch Capacity (%)	Carbohydrate Relative to N	Leaf Texture
No. 0	Yellow	Variable	Variable	Variable
No. 1	Yellow-green	75-100	High	Stiff
No. 2	Olive-green	50-75	Intermediate	Sott-stiff
No. 3	Black-green	25-50	Low	Soft

Soil nitrogen is measured as a means of helping to keep fruit nitrate levels below the 8.0 p.p.m. level considered to be critical for processed fruit. The optimum level of nitrogen in soil, based on water extraction after a 14-day incubation, is 120+ p.p.m. nitrate (NO_3), which is equivalent to 27 p.p.m. of N (Swete Kelly, 1993). The preplant nitrogen recommended is based on preplant levels of nitrate found in soil. Total plant-crop requirements for nitrogen for pineapple range from 250 to 700 kg ha^{-1} (*i.e.* 4 to 10 g plant^{-1}), depending on the soil and ecology of the site, plant population density, expected fruit mass and other environmental or management factors. Calibration experiments, leaf colour indices and tissue indices can all be used to guide growers to the correct amount of nitrogen required for optimum growth.

Under severe N deficiency, plant growth was observed to be very poor with no production of suckers, slips and fruit. Root growth was also limited, and only a few new roots were formed (Gibes and Samuels, 1958). Under adequate N nutrition, young leaves are bluish-green with a purplish midrib which becomes reddish in older leaves (Malan, 1954). Inadequate N gives a yellowish colour to leaves, with the red being more predominant at the midrib. N-deficient plants of Singapore Spanish in sand culture were found, stunted and yellow with delayed, fruiting and deformed fruits (Kanapathy, 1959). De Geus (1961) reported pale yellow to orange leaves with N deficiency, especially the order leaves. Both young and mature leaves were chlorotic. Srivastava (1963) observed stunted growth with N deficiency; emerging leaves were small and pale-green with necrotic tips. Older leaves were pale-green and had developed necrotic margins. A chemical study by Sideris (1955) showed that there was low chlorophyll and protein-N in nitrogen-deficient leaves than those receiving nitrogen. Excess N results in vigorous leaf development at the expense of fruiting.

Phosphorus

The growth of all plant parts is depressed as a result of phosphorus deficiency. However, the phosphorus requirement of pineapple is low and plants can extract P from soils having very low levels of that nutrient. The plants P requirements can almost always be met by applying P prior to planting. Soil P is the primary index used to assess the P requirement of pineapple, and levels of 20 p.p.m. or greater are adequate to sustain pineapple growth. The visual symptoms of phosphorus deficiency are not commonly seen and are not particularly specific. They can be confused with plants suffering from root injury due to such causes as drought, nematode damage or mealybug wilt. Phosphorus-deficient plants have erect, long, narrow leaves. Older leaves show leaf-tip dieback preceded by a chlorotic or red-yellow area, which extends downward along the margins of the leaves. Young leaves, primarily because of the contrast, appear to be dark green but with considerable red pigment (Swete Kelly, 1993).

Phosphorus is present in the plant not only in various organic forms such as nucleic acids, phytin, phosphatides and as constituent of enzymes, but also in inorganic forms, which actively participate in energy transport mechanism, of the plant. In all reactions of synthesis and decomposition of carbohydrates, phosphoric acid has a key role. From these reactions, plant obtains energy for its vital processes. However, pineapple absorbs very little of phosphorus, indicating that pineapple is able to use effectively little quantities of P_2O_5 it absorbs. Py *et al.* (1956) could establish no significant effects resulting from application of a phosphatic fertilizer even on a soil having relatively low phosphate contents. However, according to Nightingale (quoted by Py *et al.*, 1957a) heavy dressing of phosphorus accelerated fruit formation and ripening. Phosphorus application had shown no marked effects on fruit yield and quality (Dunsmore, 1957). Research carried out at the Indian Institute of Horticultural Research, Bangalore, revealed that phosphorus had no role in improving either fruit yield or quality (Chadha *et al.*, 1972). However, application of P was found to promote development of ratoons and had resulted in increased yields (Pan, 1957; Garcia and Treto, 1988; Mustaffa, 1989). P_2O_5 was also found to have a complementary effect on yield when applied together with N or K. Collins (1960) stated that phosphate Deficiency reduced plant vigour and production of planting-material.

With P deficiency, colour of the new leaves becomes dark-purple green without any sign of chlorosis (Nightingale, 1942; Cibes and Samuels, 1958). With the absence of phosphorus, plant growth was poor; there was no production of fruits, slips and suckers. Chlorotic areas on leaves were apparent as leaf age advanced (Cibes and Samuels, 1961). P-S resulted in reduced leaf number and growth (Srivastava, 1963). Leaf size and colour was not affected in the early stages. However, older leaves had marginal chlorosis with large bronze and purple chlorotic areas. In a sand-culture study in Malaya (Kanapathy, 1959), P deficiency resulted in dark green, purple-ribbed leaves, and fruits of such plants were sour and watery. Higher amounts of P had depressing effect on growth (Malan, 1954). Excess P depressed yield, hastened fruiting and increased number of cull-fruits (Samuels *et al.*, 1956). Yield reduction with high P was attributed to reduced N uptake (De Geus, 1961).

Mycorrhizal fungi apparently do not contribute significantly to the P nutrition of pineapple, except where soil P is extremely low, much less than 0.02 mg l^{-1} of soil solution (Aziz *et al.*, 1990), or in *in vitro* conditions (Guillernin *et al.*, 1997). In a recent study, which did not include data on P in soil or leaves, average fruit mass and yield per 12 m^2 plot were significantly greater where plants were inoculated with *Glomus mosseae* and *Glomus manihotis* (Thamsurakul *et al.*, 2000) relative to the control or to either mycorrhiza species alone. If mycorrhyzae did facilitate P uptake, they would only postpone the time when P fertilization would be required.

The leaf may not be as important as the soil as an index for P, but leaf analysis provides a check on the adequacy of soil supplies. Leaf P in the 'D'-leaf basal white tissue naturally increases with age–for instance, from 0.01 per cent (100 p.p.m, in 'D'-leaf basal white tissue on a fresh-mass basis) at 5 months to about 0-02 per cent (200 p.p.m.) at flower induction (Swete Kelly, 1993). According to Py *et al.* (1987), leaf P in the whole 'D' leaf should be 0.8 per cent of dry matter at the time of flower induction.

Potassium

Potassium is found in inorganic forms everywhere in the plant; where most of the physiological activities are in progress, particularly, in young leaves and meristematic tissues. Formation of sugar and starch, of organic acids, and of hardened tissues, reduction of nitrates and consequently synthesis of proteins are largely dependent on the quantities of potassium present. Potassium is indispensable in transportation of photoassimilates from leaves to organs of storage and to organs of utilization in the physiological processes. K is involved in stomatal regulation and water economy also.

A copious supply of potassium increased stem weight to a greater extent than the weight of the plant, leading to the development of a strong stem with higher accumulation of starch and protein (Sideris and Young, 1945). Pineapple has high potash requirement soon after planting, and its requirement is highest at the time of fruit formation and ripening. Potassium has been reported as the factor controlling fruit quality (Py *et al.*, 1957a). Fruits from potash-deficient plants were small with poor quality having low sugar and acid contents (Kanapathy, 1959). Diameter of fruit-stalk and the number of fruits standing erect on the plant were found more when the proportion of potassium was high in the total cations applied (Martin-Prevel, 1961).

Potassium deficiency leads to appearance of brown spots on green parts of the leaves and withering of leaf tips. The leaves are also shorter and narrower than normally nourished leaves (Py *et al.*, 1956). Initially, K-deficient leaves remain green with only leaf-tip drying, and necrotic spots are formed on blistered areas which appear later on leaf surfaces. These symptoms appear first on older leaves and progress to younger ones (Cibes and Samuels, 1958). Fruits also from K-deficient plants were small, late maturing, low in total solids and acids (Cibes and Samuels, 1961). Martin-Prevel (1959b) found senescent and recently matured leaves to drop and develop numerous small translucent yellow spots (3 mm diameter). These symptoms were more pronounced as leaves aged. Spots appeared as raised areas on the upper surface and later spread over the entire leaf, excepting the basal 15 cm. They coalesced into large yellow spots and finally the center of the yellow spots became necrotic. Leaves of K-deficient plants were markedly smaller and had necrotic tips and margins (Srivastava, 1963). Absorption of potassium continues actively even after flower induction (Lacoeuilhe, 1973), and represents double the mass of nitrogen in the plant until flower induction (Lacoeuilhe, 1978).

Potassium, like nitrogen, is required in large amounts to sustain pineapple plant growth. Potassium deficiency would decrease photosynthesis and thus plant growth, fruit mass and slip production (Swete Kelly, 1993). With potassium deficiency, fruits have reduced sugar and acid levels and have a pale colour (Py *et al.*, 1987; Swete Kelly, 1993), presumably because of reduced carotenoid development. Where K is defiant, the fruit peduncle diameter is reduced (Py *et al.*, 1987), the peduncle is weak (Swete Kelly, 1993) and fruits are more prone to lodging and sunburn and have lower acidity aid aroma development.

As with phosphorus, the primary index for K is the level in the soil, because K is well retained by most soils. The optimum soil level at planting is 150 p.p.m. and potassium deficiency symptoms are observed when the soil level is below 60 p.p.m. (Swete Kelly, 1993).

The high requirement of pineapple for potassium has made it relatively easy to reproduce deficiency symptoms. Low levels of potassium are associated with shorter leaves that are narrower in relation to their length, growth is reduced, necrotic spots can be seen in the green photosynthetic tissue (chlorenchyma) of the leaves and leaf tips die back (Py *et al.*, 1987; Swete Kelly, 1993). During the early stages of potassium deficiency, leaves are dark-green and narrow, but, if the deficiency is prolonged, leaves eventually become yellow.

Potassium analysis is done on the basal white tissue of 'D' leaves (Swete Kelly, 1993) or whole 'D'-leaf samples (Py *et al.*, 1987). Visual potassium deficiency symptoms are evident when there is less than *0.20 per cent* K (2000 p.p.m.), fresh-mass basis (2.4 per cent on a dry-mass basis), in the basal white tissue of 'D' leaves. The critical leaf K level at flower induction is reported to be, 0.30 per cent (3000 p.p.m.) on a fresh-mass basis.(3.6 per cent on a dry-mass basis) for basal white tissues (Swete Kelly, 1993;) or 2.2-3 per cent for the whole 'D' leaf on a dry mass basis (Dalldorf and Langenegger, 1976; Py *et al.*, 1987). Swete Kelly (1993) recommends that leaf K of plants 3-5 months old should range between 3500 and 4000 p.p.m. (basal white, fresh-mass basis).

Calcium

Like all other plants, pineapple also needs calcium as it is indispensable for cell-wall formation and for cell division. Accumulation of calcium in leaves is characteristic of aging and maturation. Though calcium needs of pineapple are small, but an excess of cations in the soil leads to calcium chlorosis.

Calcium is distributed uniformly over the whole leaf in pineapple; whether it is chlorophyllous or non-chlorophyllous tissue of the leaf (Sideris and Young, 1945). Calcium accumulation in stem is twice as much as that in leaves. It accumulates especially in meristematic tissues of the stem apex. With the formation of placenta and fruit, calcium content of stem gets reduced (Sideris and Young, 1951). Calcium seems to play an important part in the flower-bud differentiation and in fruit formation (Sideris and Young, 1950).

Calcium deficiency is observed in young leaves; as calcium is immobile in plant. Calcium deficiency leads to pale-green leaves with some yellow mottling and new leaves show dieback of tips (Cibes and Samuels, 1958). As deficiency progresses, leaves develop blisters and a reddish colour at the base. Fruits with calcium deficiency have blackish spots internally with a gelatinous appearance. Kanapathy (1959) found that with calcium deficiency leaves were soft, flat and dark green and fruits were tasteless.   Excess calcium resulted in partially chlorotic plants with stunted growth (Collins, 1960), as it inhibits absorption of iron.

Pineapple has a very low requirement for calcium, but deficiencies can occur on highly weathered soils low in basic cations and on soils where pH has been lowered by long-term use of acidifying fertilizers, such as ammonium sulphate. For optimum growth, soils should contain greater than 100 p.p.m. of Ca, a level approximately one-tenth that normally recommended for most crops, and deficiency symptoms are observed when the level is less than 25 p.p.m. Calcium is commonly applied to amend soil pH as well as to supply Ca, but, in many areas, the soil pH is kept at or below 5.5 to limit the incidence of heart and root rots caused by *Phytophthora* spp. Since the amount of calcium required to adjust pH varies with soil cation exchange capacity, the lime requirement should be based on a lime titration curve developed specifically for each soil type in which pineapple is grown. Gypsum can be used where it is desirable to supply calcium without changing soil pH. However, only one reference (Hartung *et al.,* 1931) was found indicating that gypsum might have been evaluated as a source of calcium for pineapple.

Leaf colour of calcium-deficient plants was abnormal, a grimy grey-green rather than the more normal clean yellow-green or green. The initial growth of calcium-deficient pineapple plants may not be stunted, but as the deficiency becomes more severe, growth depression is very evident.

Calcium deficiency symptoms, as with those of boron, are most likely to be seen initially on the fruit because the demand for both calcium and boron in the growing point is highest at the time of floral differentiation. When Ca deficiency is severe, cells at the growing point fail to divide and other cells tear apart because of weak cell walls, so new leaves may appear to be cut off or tipless, with serrations or scalloping along the margins, and leaves may develop callus, be abnormally thick and have streaks of corky tissue running parallel to their length. With extreme deficiency, the growing point may die, resulting in the growth of side-shoots, which may be initially symptom-free. In some cases, plants fail to produce an inflorescence and continue to grow vegetatively. In this case, the leaves become progressively shorter as they develop. Roots are thicker than normal and, as a result, such plants are more difficult to pull out of the ground than a normal plant. Fruits may be abnormal in size and shape. Symptoms have been observed only for plants having a concentration of 0.002 per cent or less, fresh-mass basis (0.024 per cent, dry-mass basis), in the basal white tissue of the 'D' leaf, The critical concentration at flower induction is 0.015 per cent fresh mass (0.18 per cent, dry-mass basis) in the basal white tissue, and deficiency symptoms develop when the level is less than 0.004 per cent (Swete Kelly, 1993). In the whole 'D' leaf, leaf Ca should be 0,10 per cent of dry matter at the time of induction (Py *et al.*, 1987).

Magnesium

Magnesium is an important component of chlorophyll, green-colouring substance of the leaf, and is absolutely essential for any green plant. Though Mg requirement of pineapple is not large, its deficiency leads to restricted formation of chlorophyll, causing reduction in photoassimilates. Besides, Mg is required for activation of several enzymes involved in photosynthesis and respiration.

Unlike calcium, magnesium accumulates more in leaves than in stem (Sideris and Young, 1946). Magnesium plays an important role in increasing average fruit weight as well as yield (Martin-Prevel, 1961

Chlorosis becomes evident first in older leaves with magnesium deficiency as it is mobile. Under severe deficiency, yellow mottling appears, which at times coalesces to form an almost yellow stripe along the leaf margin; and leaves have a reddish tinge (Cibes and Samuels, 1958). Kanapathy (1959) noted pale-greenish yellow leaves. Fruits were sour without firm flesh with Mg deficiency. Plants with Mg deficiency have poor root anchorage (Glennie, 1977).

Magnesium is a component of the chlorophyll molecule, and a deficiency will reduce chlorophyll concentration, photosynthesis and growth. This nutrient is mobile in the plant and the predominant visual symptom of magnesium deficiency is bright yellow older leaves, particularly those leaves or parts of leaves exposed to sunlight. Such leaves will frequently have bands of green that run diagonally across the leaf as a result of being shaded by leaves above them (Py *et al.*, 1987; Swete Kelly, 1993). The stems of Mg-deficient plants are short and have a small diameter. The root systems tend to be weak, so magnesium-deficient plants are easily pulled from the soil. Fruits are reported to be low in acidity, sugar content and aroma (Py *et al.*, 1987). This symptom probably reflects the plant's reduced capacity to assimilate CO_2 via photosynthesis.

At planting time, the optimum level of Mg in soil is 50 p.p.m. and Mg deficiency occurs at levels below 10 p.p.m. (Swete Kelly, 1993). Swete Kelly (I993) states that deficiency symptoms begin to develop when Mg in the basal white tissue of 'D' leaves reaches 0.015 per cent fresh mass (0.18 per cent, dry-mass basis) and symptoms are present when the Mg content in the basal white tissue of 'D'-leaves is 0.00970 or less, fresh-mass basis (0.108 per cent, dry-mass basis). The critical concentration in fresh 'D'-leaf basal tissue at floral induction is reported to vary from 0.0257 per cent (in low-potassium soils) to 0.027 per cent (in high-potassium soils) (0.30 per cent and 0.32 per cent, respectively, dry basis) (Swete Kelly, 1993). Magnesium in the whole 'D' leaf should be 0.1870 of dry matter at the time of induction (Py *et al.*, 1987).

Sulphur

Sulphur, being a constituent of some amino acids aid all proteins, is very much needed for plants. Its requirements by tie pineapple-plant are as high as that of the phosphorus (Teiwes and Gruneberg, 1963). The deficiency of sulphur is a rare occurrence since sufficient quantities of this nutrient reach the soil as constituent of fertilizers such as ammonium sulphate, superphosphate as well as sulphate of potash. In addition, sufficient quantity of sulphur is carried by rains from the atmosphere to soil; especially in the industrial areas.

In the early stages of sulphur deficiency, leaves show very little change in appearance except for some blistering on older leaves. As deficiency advances, leaves turn lighter green, as happens in mild N deficiency. Fruits produced from S-deficient plants of Red Spanish variety exhibit abnormalities in structure and ripening. Unlike normal fruits, which ripen from base to top, S-deficient fruits ripen from top to base. A vacuole is also observed inside the fruit at the junction of ripe and unripe portions (Cibes and Samuels, 1958). Plant and fruit growth and slip and sucker production remain normal, excepting of the above mentioned fruit-development abnormalities (Cibes and Samuels, 1961). Of all the elements, sulphur deficiency has very little effect on yield.

Sulphur deficiency is rare, probably due to the fact that many fertilizers contain sulphates. Deficient plants have bright lemon-yellow leaves that are broader than normal. As contrasted with magnesium deficiency, where chlorosis occurs mainly on older leaves, both young and old leaves of sulphur-deficient vegetative plants are yellow. As the deficiency progresses, later-formed leaves become narrow, plants are stunted and fruit size is reduced (Py *et al.*, 1987). The symptoms described above were associated with a sulphur level of 0.005 per cent, fresh-mass basis (0.06 per cent on a dry-mass basis) in the middlethird of 'D' leaves.

5.5 Micronutrients

On the basis of the frequency with which typical symptoms of deficiency occur in pineapple, iron can be regarded as one of the most important micronutrients, followed by zinc, copper, boron and manganese.

Iron

Iron, being a constituent of several enzymes, plays an important role as redox catalyst, participating in oxidation of cabohyorates and in reduction of sulphates and nitrates (Sanford *et al.*, 1954). Though iron is not a component of chlorophyll, its deficiency leads to reduced synthesis of the pigment. Iron availability is very much limited in calcareous soils and in soils with high pH. Not only chlorophyll content but also protein content diminishes with iron deficiency (Sideris, 1948). Fresh weight of plant as well as fruit decreased in the absence of iron. The absorption and translocation of iron hinders within the plant with the presence of phosphates (Sideris and Young, 1956). Calcium chlorosis (Malan, 1954) occurring in pineapple in calcareous soils is on account of lime-induced iron deficiency.

The typical symptom of iron deficiency is chlorosis, disappearance of green, leaf-colouring substance, which also occurs with N deficiency. Chlorosis caused by iron deficiency appears on younger leaves in rosette (Sideris, 1955), and the one caused by N deficiency occurs first in older leaves (Py *et al.*, 1957a). In nitrogen deficiency, leaves formed before and during the period of deficiency become chlorotic in varying degrees. In iron deficiency, only the leaves, which are formed during the deficiency period become chlorotic, while those formed before this period remain relatively green (Sideris and Young, 1956). Though mild N and Fe deficiencies are similar in appearance, the Fe deficiency normally differs from N with the presence of reddish tinge on the whole leaf (Cibes and Samuels, 1958).

In Hawaii and Puerto Rico, regular spraying of ferrous sulphate forms part of the fertilizer programme of many plantations (Py *et al.*, 1957 a). In Hawaii, young plants are sprayed with ferrous sulphate at intervals of 14 days, whereas old plants are sprayed at monthly intervals (Collins, 1960).

Iron is an immobile nutrient so iron-deficiency symptoms always appear first on young leaves. In Hawaii, a visual index was developed based on the percentage of the leaf area ('B' through 'F leaves only). Total iron in the middle third of 'D'-Ieaf green tissue expressed on a fresh-mass basis was also used in Hawaii as an index for determining iron requirements. Where the 'D'-leaf level is 8 p.p.m. or higher in the absence of high levels of soluble manganese, visual iron deficiency is 40 per cent or lower, a level considered adequate for maximum growth. In the presence of soluble soil Mn, the iron content of leaves is not well related to the existence of deficiency symptoms. Also, unlike the situation with other nutrients, pineapple plants can show visual symptoms of iron deficiency with no decrease in yield.

Where soluble manganese is high–a common situation in soils having high manganese and low pH–the ratio of soluble iron to soluble manganese in the soil and in the whole 'D' leaf is more important than the absolute amounts of either element (Hopkins *et al.*, 1944). Py *et al.* (1987) report that iron deficiency occurs and visual symptoms are observed where iron in the 'D' leaf is between 60 and 475 p.p.m., dry-mass basis, and the Fe:Mn ratio is less than 0.4. Manganese deficiency occurs and visual symptoms are observed if Mn in the 'D' leaf is between 29 and 78 p.p.m., dry-mass basis, and the Fe: Mn ratio is greater than 10.5. Iron deficiency has been observed in Queensland, Australia, when cold, wet soils prevent the uptake of iron or where roots have been damaged by pests (Swete Kelly, 1993).

The initial initial symptoms appear as interveinal chlorosis of the younger leaves; leaf veins, which run parallel to the length of the leaf, remain green whereas the areas between the veins are yellow-green or yellow. When the deficiency is mild, the leaves become yellow with green mottling (Swete Kelly, 1993). As the deficiency becomes more severe, the entire surface of the leaves may be pale yellow or creamy white, with considerable red pigmentation at the terminal ends. Such leaves are also soft and leathery, rather than rigid, and have considerable tip dieback. Plants with severe iron deficiency will have fruit that are small, hard and reddish in colour and with cracking between the fruitlets. The crowns will be light yellow or creamy white in colour. In the absence of high soluble manganese levels, severe deficiency symptoms have been associated with 3.0 p.p.m. or less of iron, on a fresh-mass basis (36 p.p.m. on a dry-mass basis), in the middle third of 'D' leaves. As noted above, with high levels of soluble manganese, the ratio of iron to manganese is more critical than the absolute amounts of either element. Iron sulphate sprays, often applied biweekly, are used to correct the deficiency. To be effective, the iron in iron sulphate sprays must be applied in reduced form, and good storage conditions are required to prevent oxidation.

Zinc

Zinc is required for synthesis of tryptophan, which is a precursor for indole acetic acid (IAA) synthesis. It is involved in the activation of several enzymes such as alcohol dehydrogenase, lactic acid dehydrogenase, glutamic dehydrogenase and carbonic anhydrase.

Zinc deficiency is often found in peaty and sandy soils, which have high leaching rate or are strongly acidic with low pH (Aldridge, 1960). It can also occur on well-drained soils with a low humus content, if these soils have been strongly leached.

Smaller or larger transparent yellow spots on the leaf-lamina are characteristic of zinc deficiency (Tisseau, 1959). First symptoms of zinc deficiency are curling and twisting of leaves inside the rosette. Later leaves become narrow, acquire light green to yellow colour and their surface is covered with a thick layer of wax (Aldridge, 1960). In acute deficiency, inner twisted leaves bend horizontally and downwards in bundles, giving plant a resemblance with a calabash fruit. This deformation is named as Crook-neck disease and is caused by a physiological disturbance due to zinc deficiency. Another characteristic of zinc deficiency is its occurrence in scattered spots in the field-in which apparently healthy plants stand directly along the side of the deficient plants (Dunsmore, 1957). Rehm (1956) observed chlorosis in Swaziland (South Africa) without any leaf curling.

Tisseau (1959) recommended a foliar spray of 1 per cent zinc sulphate solution for correcting Crook-neck in Guinea. A spray solution of 2,000 litres is sufficient for treatment of plants in one hectare, *i.e.* 20 kg of zinc sulphate per hectare. With very acute zinc deficiency, spraying should be repeated. Zinc solution at 2 ppm improved fruit quality and 3 ppm was found to promote root growth (Srivastava, 1969).

Zinc deficiency can occur in soil with a pH of 6.0 or higher, with low organic-matter content (observed in Hawaii in such conditions) or where lime or phosphorus were not well incorporated or were applied in excessive amounts (calcium- or phosphorus-induced zinc deficiency). Zinc deficiency is widespread in Queensland, Australia (Swete Kelly, 1993), especially on previously uncultivated land. The deficiency has also been observed in Hawaiian soils that have low native fertility. When the deficiency is severe, the plant's central cluster of leaves is curved (crookneck), especially with younger plants (Swete Kelly, 1993). When the deficiency develops in older plants, the surfaces of the leaves develop yellowish-brown, blister-like (elevated) spots. The centre leaves may on occasion have rips or serrations on their edges. In less severe cases, the blisters occur only on the older leaves and the centre leaves are only slightly curved. Occasionally, the curved leaves will be seen without blisters. At times, zinc-deficient plants, like calcium-deficient plants, have been observed to remain continuously vegetative. Zinc deficiency is distinguished from calcium deficiency by the curved central leaves and the presence of blisters on the leaf surfaces. Zinc concentration in the 'D' leaf is not diagnostic of zinc deficiency. A level of 4 p.p.m., fresh-mass basis, in the stem apex is considered adequate and 3.0 p.p.m. or less, fresh-mass basis (36 p.p.m. on a dry mass basis), in the stem apex will be associated with typical zinc-deficiency symptoms. The deficiency is easily correctable with sprays of zinc sulphate.

Boron

Boron acts as a regulator of permeability of plasma and promotes absorption of cations, but represses absorption of anions. Deficiency of boron leads to collapse of conducting vessels such as cambium and phloem, which serve for transport of photoassimilates from leaf to storage organs. Cambium and phloem suffer to a greater extent than xylem vessels; conducting water and nutrients. The consequence of inadequate boron supply is accumulation of products of assimilation, reducing sugars and soluble nitrogen compounds, in leaves.

No visual symptoms on leaves are seen due to boron deficiency. However, there is a visible influence on fruit production; as fruits produced are abnormal in size and shape. The fruit is also small with pronounced separation and cracking of fruitlets. The cracks between fruitlets are filled with gum (Cibes and Samuels, 1958). Glennie (1977) reported brown corky flecks in inter-fruitlet areas. In severe cases, plants produce corky cricket-ball-size fruits or no fruits

Fruitlets of immature fruit showing boron deficiency appear glossy and green in contrast to the scurfy, dull and whitish appearance of a normal fruit *at this* stage. The glossy appearance is due to the absence of trichomes (multkellular plant hairs). As the fruit develops, small, shallow cracks appear between the fruitlets and within 2 weeks these cracks become corky Fruit borne on plants having severe boron deficiency are much smaller than normal fruit, and multiple crowns are also reported (Py *et al.*, 1987).

The symptom of boron deficiency on vegetative leaves, if it occurs, is death of the tips of the youngest leaves, sometimes with serrations on the margins. In extreme cases, death of the growing point will occur. In such a plant the central leaves will be stiffer and shorter than normal and lateral buds will eventually develop. The symptoms of boron and calcium deficiency on vegetative plants are similar. As in the case of calcium, the initial symptoms of boron deficiency in the field are most likely to be observed on the fruit rather than on vegetative plants, because the greatest demand for both boron and calcium is at the growing point as it shifts from the production of vegetative structures to reproductive ones. Symptoms of boron deficiency can be expected to occur when boron is 0.2 p.p.m. or less, on a fresh-mass basis (2.4 p.p-m. on a dry-mass basis), in the middle third of 'D' leaves approximately 10 months after planting. Fruit symptoms were associated with a level of 0.4 p.p.m. or less boron in the middle third of the longest crown leaves 4.5 months after floral differentiation. In Australia, boron deficiency is prevented by forcing of flower induction with sprays that contain 0.5 per cent borax. Borax provides a source of B as well as raising the pH of the ethephon solution to about 9.0 to enhance its effectiveness, especially when warm temperatures make induction difficult (Sinclair, 1994).

Manganese

Manganese deficiency is rare and occurs in soils high in calcium with *a* high pH. Pyet al. (*1987*) report that manganese deficiency symptoms are not specific. Affected leaves are marbled with pale green areas, mainly where vessels are located. Despite the presence of high levels of soluble manganese in many tropical soils, including those in Hawaii, symptoms of manganese toxicity have not been observed. Pineapple growing in acid soils tolerates high levels of both soluble manganese and aluminium where other plants show symptoms of toxicity. In acid, high-manganese soils, high levels of soluble manganese appear to interfere with iron absorption and translocation (Sideris 1950) or utilization. As noted above, the iron-manganese ratio is more important than the absolute amount of either element.

Copper

High concentration of copper to an extent of 70 per cent occurs in leaf-chloroplast. Copper activates several enzymes, *viz.* tyrosinase, cytochrome oxidase, polyphenol oxidase, ascorbic acid oxidase and phenolases.

Crook-neck disease is often identified with copper deficiency (Aldridge, 1960; Tisseau, 1959). A physiological disturbance called green die-back was found in Malaya due to copper deficiency alone (Teiwes and Gruneberg, 1963). It is characterized by leaves, which are thinner and narrower than those of healthy plants. Younger leaves of Cu-deficient plants are shorter and narrower. Severe copper deficiency causes death of plants.

Copper sulphate should not be applied as spray to leaves since it causes severe leaf scorching (Tisseau, 1959; Srivastava, 1969). Spraying the ground in the neighbourhood of the plants with 1,5-2.0 per cent solution is recommended; 30-50 ml of the solution is distributed at a distance of 3-5 cm from the base of the plant (Tisseau, 1959). Since copper sulphate has a strong corrosive action, the use of metallic vessels should be avoided as far as possible. In replanting plots on which Crook-neck has already occurred, a preventive soil treatment should be carried out by mixing 25-40 kg of copper sulphate with fertilizers to be applied per hectare as a basal dressing. Annual dressings of 6-11 kg/ha are recommended along with fertilizer mixture to avoid green die-back in Malaya.

The leaves of plants deficient in copper are lighter green than those of normal plants and are distinctly U-shaped in cross-section relative to normal leaves. Tips of leaves curve downward instead of being erect. The deficiency is common in the heavily leached sandy soils of southern Queensland, Australia (Swete Kelly, 1993), and has also been observed in Malaysia on peat soils. The optimum range in the 'DMeaf basal white tissue is 10-50 p.p.m. on a fresh-mass basis. Copper deficiency is easily corrected with a copper sulphate spray.

Molybdenum

There are no known reports of visual symptoms of molybdenum deficiency on pineapple and also little indication of a pineapple growth response to Mo. Molybdenum is essential for proper functioning of the nitrate reductase enzyme, and it was reported that application of Mo can reduce the nitrate level in fruit juice (Chairidchai, 2000). Much additional work in Queensland, Australia, has failed to demonstrate any change in juice nitrate levels as a result of spraying plants with Mo (Scott, 2000).

Deficiencies of the macronutrients N, P, K, Ca and Mg are likely to occur anywhere that pineapple is grown its quantities removed by the crop and lost by leaching are not replaced through fertilization. Deficiencies of S and of the micronutrients Fe, Zn, B, Mn, Cu, Mo and Cl are likely to be localized in specific areas where pineapple is grown. The sulphur requirement of pineapple, which probably is as high as that for P, will be met by S-containing fertilizers if iron, potassium and zinc sulphate are applied to pineapple fields, though this situation could change if the application of these fertilizers is reduced or eliminated. Since the main function of Mo in higher plants is related to nitrate reduction, there is little chance that a Mo deficiency will occur in pineapple in most regions because most N taken up by pineapple is in the form of ammonium or urea, both of which can be absorbed through the leaves. However, Mo deficiency, at least in terms of insufficiency to reduce plant-absorbed nitrate, has been reported in Thailand (Chongpraditnun *et al.,* 2000). Boron and copper deficiency are already found in a few countries and these two nutrients may eventually become limiting in other pineapple-growing regions. Iron deficiency is common in many pineapple growing regions while zinc deficiency seems to be somewhat less widespread.

As long as fertilizers are readily available and inexpensive relative to the value of the crop and the technology is available to apply these fertilizers after the deficiencies appear, deficiencies of most nutrients should not limit the productivity of pineapple where the crop is grown for commercial purposes. This is because most deficiencies are readily correctable by foliar application of soluble sources of nutrients. The pineapple plant is ideally suited for foliar fertilizations and most nutrients are readily absorbed through the leaves or are taken up when solutions containing essential nutrients flow to the leaf axils, where roots commonly exist to absorb them.

5.6 Plant Requirements/Nutrients Removed by the Crop

Nutrients are exported from the field in fruits and crowns, slips, suckers and other propagules harvested for planting material. Hence, the amounts of exported nutrients are related to fruit and propagule yields. Mineral contents of fruits, in percent, are reported to be in the range 0.075-0.08 N, 0.015 P.O., 0.2-0.26 K_2O, 0.015-0.02 CaO and 0.13-0.IS MgO on a fresh-mass basis (Py *et al.,* 1957).

Crop Residue

Significant vegetative residue remains at the end of the pineapple cycle and the amount and the quantity of nutrients it contains is determined, at least in part, by the quantity of fertilizer applied during the crop cycle. This residue can vary from about 70 to 240 t ha^{-1} on a fresh-mass basis or 40-60 t ha^{-1} on a dry-mass basis (Py *et al.*, 1987; Pena Arderi and Dominguez Martin, 1988). Pineapple residue is approximately 75 per cent leaves and 25 per cent stems, with the root system representing a relatively insignificant amount. Most of the dry residue consists of organic compounds, such as sugars, starch, cellulose and proteins. The residue contains, in per cent, approximately 1.0 N, 1.0-2.0 K and approximately 0.1-0.4 P, Ca, Mg and S and much smaller amounts of Fe, Zn, B, Cu, Mn and Mo, though values can range rather widely at the end of the crop cycle (Table 8). The incorporation of the residue could return substantial amounts of organic matter to the soil, as well as contributing significant amounts of nutrients to a subsequent crop of pineapple (Table 9). Organic farming using pineapple ratoon compost pineapple ratoon compost improves the soil condition and fertility. Organic fertilization of pineapple with its own ratoon compost-based on current N doses needs to be standardized so that there should not be shortage of nutrients and reduction of yield (Singh 2010).

The supply of plant nutrients from the residue to a subsequent pineapple crop depends on how the residue is handled. Chopped green tops of pineapple are a good source of roughage for ruminant animals (Henke, 1934; Anon., 1944; Bishop *et al.*, 1965; Wayman *et al.*, 1976; Py,

**Table 8: Amounts of Nutrients Removed by Crowns, Fruits
and Suckers (Martin-Prevel, 1961a,b,c,d;
Martin-Prevel *et al.*, 1961; Py *et al.*, 1987).**

Organ	Nutrients (kg ha⁻¹)					
	Fresh Mass (g)	*N*	*P_2O_5*	*K_2O*	*CaO*	*MgO*
Crown*	205	19	11.5	59	8.5	10.5
Crown*	295	28	15.4	89	11.9	13.0
Crown*	390	36	20.2	111	29.3	16.5
Fruit++		43	16.5	131	17.0	10.0
Sucker ++		24.5	8.0	43	10.0	6.2

*: Cote d'Ivoire, 5.5 plants m⁻²,+ Martinique, 5.5 palnts m⁻²,
++ 3.8 plants m⁻².

1978) and, if harvested for this use, will also remove large
amounts of plant nutrients from the field. Since most of the
nutrients accumulate in the green leaves, the nutrients lost
in harvested leaves would be considerably greater than the
fraction of plant mass removed. Burning the residue rather
than incorporating it into the soil will volatilize most of the
N and S, while all other nutrients will be converted to soluble
inorganic forms. Nutrients in soluble form are readily
available to a subsequent crop, but are also subject to
leaching by rainfall.

Pineapple residue that was chopped and then
incorporated into the soil decomposed in approximately
26 weeks in Hawaii. Early in the decomposition process,
much of the N and probably some of the S and P would be
utilized by microorganisms involved in the residue
decomposition process. Consequently, the available N and
P decreased during the period of rapid residue
decomposition. On the other hand, available K, which is

Table 9: Inorganic Nutrient Content of Pineapple Plant Residue at the End of the Crop Cycle

Element	Concentration (% dry weight)	Average Nutrient Content in a Given Amount of residue (kg ha^{-1})		
		30 t ha^{-1}	50 t ha^{-1}	70 t ha^{-1}
Wayman *et al.*, 1976				
N	0.78± 0.13	234	390	546
K	0.94±0.43	282	470	658
P	0.07±0.01	21	35	49
Ca	0.43±0.21	129	215	301
Mg	0.16±0.03	48	80	112
S	0.23±0.09	69	115	161
Mn	0.05±0.01	15	25	35
Fe	0.04±0.01	12	20	28
Cu	0.003±0.001	0.9	1.5	2.1
Si	0.32±0.19	96	160	224
Na	0.01±0.01	3	5	7
Cl	0.36±0.05	108	180	252
Al	0.13±0.08	39	65	91
Ingamells, 1981				
N	1.03	309	515	721
K	1.94	582	960	1358
P	0.19	57	95	133
Ca	0.31	93	155	217
Mg	0.19	57	95	133
S	0.07	21	35	49
Mn	0.027	8.1	13.5	18.9
Fe	0.018	5.4	9	12.6
Cu	0.001	0.3	0.5	0.7
Si	0.003	0.9	1.5	2.1
Na	0.31	93	155	217
Cl	0.08	24	40	56
Al	0.029	8	14	20

primarily present in free (unbound) form in plant tissue, steadily increased during the decomposition period.

The elements Ca and Mg, which are primarily in organic form in the residue, are released later than K. Ingamells (1981), who compared the growth of pineapple in four soil series with and without pineapple residue, reported that the incorporation of residue into the soil immobilized N for 30-60 days, after which its availability increased. In the Hawaii environment, this initial reduced availability would be of little consequence, because roots do not begin to proliferate through the soil until 30 or more days after planting. The change in availability *of N* had no significant effect on the concentration of the X in the pineapple plants grown in the soil. At 16 weeks after planting, plants grown in soil in which residue had been incorporated had higher concentrations of K, Cl, Ca, Si, Mg, P and S in pineapple leaves than did plants grown in soil with no residue (Ingamells, 1981). Since the incorporation of pineapple residue into the soil also increased the moisture-holding capacity of the soil, the increase in the level of these elements in plant leaves may be due to the combination of a higher level of these nutrients and greater available soil moisture.

The benefits from the incorporation of residue into the soil are difficult to evaluate, especially for a long-term crop like pineapple grown in the tropics, where there is ample time during the crop cycle for complete decomposition of the residue. As with any incorporated organic matter, pineapple residue can improve soil physical, chemical and biological characteristics, but a direct effect on the following crop may not be obvious. In a study of residue management on a ferralitic sandy soil in Cote dTvoire, residue was incorporated, burned or mulched (Godefroy, 1979). The plants were spray-fertilized with N (amount not given) and 1000 kg ha^{-1} of K$_2$O. Nitrogen losses where high– approximately 200 kg ha^{-1} in all treatments. Potassium losses due to leaching were lower than for nitrogen, averaging, in kg ha^{-1}, 189 where residue was incorporated, 137 where it was burned and 107 where it was mulched.

5.7 Nutrient absorption and growth

Pineapple plant nitrogen and potassium requirements are low until about 4 months after planting (Lacoeuilhe, 1978; Ingamells, 1981), after which requirements increase with growth until flower induction. In an experiment where 8 g of N and 20 g of K$_2$O were provided for each plant prior to forcing, at the time of forcing at 10 months after planting plants of 375 g dry mass contained 5 g N and 11 g K (Lacoeuilhe, 1978). Plant nitrogen content remained constant during the period from forcing until harvest, However, K absorption continued to increase from forcing to harvest, and 13 g were, accumulated in a plant dry mass of 800 g (Lacoeuilhe, ' 1978). Factors that reduce potential growth, such as drought, low temperature and root anoxia, reduce the plant's nutrient requirement (Py *et al*, 1987).

Total uptake of a particular nutrient does not necessarily indicate a plant's requirements for that nutrient. Pineapple plants may take up more, the same as or less than the amounts required for optimum growth, depending on the factors indicated above and the specific nutrient in question. In a field high in available N and K, total uptake over a 30-month period was highest for K, followed by N and phosphorus. The efficiency of pineapple in extracting K from the soil is high and, if readily available, the plant accumulates K in greater amounts than are required for optimum growth, often referred to as luxury consumption.

The uptake of nitrogen also shows luxury consumption, generally being proportional to the amount of nitrogen fertilizer applied during vegetative growth (Scott, 1993). Juice in nitrate is an important quality factor in canned pineapple because high levels detin cans; 8 p.p.m. is considered the critical level in Australia (Scott, 1994). Juice nitrate levels can highly correlated with nitrogen applied with fertilizers (Scott 1993). In one study juice nitrate averaged 1.0 6.0 and 23 p.p.m. N applied 200, 600 and 1200 kg ha^{-1} (Scott 1993). Heavy N fertilization and application of N fertilizer after flowers is more likely to result in elevated fruit nitrate levels (Chongpradium *et al.,* 1996). However, nitrate levels in leaves sampled at different stages of plant growth were not well correlated with juice nitrate levels (Scott, 1994). Fruit from plants with a low level of molybdenum in the 'D' leaf had a high nitrate content in juice in Thailand Chairdchiai, 2000; Chongpradium *et al.,* 2000). In these experiments, high fruit nitrate content and low 'D' leaf molybdenum content were thought to be due to increased absorption of molybedenum by soil particles at the low soil pH.

Luxury consumption of calcium (Godefroy *et al.,* 1971), magnesium sulphur, born, chlorine copper and maganese also occurs in pinepalle when these elements are readily available, whether they are taken up from the soil or applied to the leaves in nutrient solutions (WG Sanford, personal communcation). Conversely low amounts of phosphorus are extracted by the plant and uptake is not proportional to the supply available, but reflects the plants requirement for P. The levels of iron zinc and molybdenum also do not generally increase with the available supply, but reflects the plants requirements for these nutrients. However, as was noted above, leaf/leaves of iron and manganese can be above 500 mg kg^{-1} dry mass in soils that have a low pH (3.5-4.5) and high levels of soluble manganese.

5.8 Effect of Nutrition on Fruit Quality

Assuming that no other factors limit growth, the adequacy of the nutrient supply determines the plant growth rate, the plant mass at induction and ultimately the fruit mass at harvest. The literature seems to indicate that plants well supplied with nutrients at the time of induction are likely to have larger fruit than plants of the same mass that have less than optimum nutrition (Py *et al.*, 1987). If nutrition is adequate at the time of induction, additional nutrients are typically not applied after that time, because nutrient absorption, except for potassium, ceases (Py *et al.*, 1987). Fruit mass is well correlated with plant mass at induction, and fruit quality is primarily determined by environmental factors. Where nutrient supplies at the time of induction are inadequate, fruit mass may be increased and fruit quality can be affected by the application of fertilizers after induction has occurred. Since fruitlet numbers are fixed soon after induction, fruit mass is only increased to the extent that additional nutrients result in an increase in fruitlet size. The focus of the following discussion relates mainly to the effect of nutrition on fruit quality.

According to Py *et al.* (1987), N and K are the most important elements influencing fruit mass and quality in relation to each other and in relation to climate. It is not always possible to distinguish between the specific effect of N on fruit quality and its more general effect on overall plant growth and health. Hence, an increase in N increases the diameters of the core and the peduncle and the length of the peduncle. As a result, with soils and losses by teaching do not represent an environmental hazard. The fate of the large amount of N applied to pineapple remains insufficiently known and more work should be done to understand the nitrogen dynamics of pineapple-based cropping systems. Studies are needed to demonstrate that the large amount of nitrogen applied to the well-managed pineapple crop does not contaminate groundwaters with nitrate. It is likely that environmental concerns will eventually require that farmers manage nutrients in a way that protects both ground- and surface waters from nutrient contamination.

Vegetative growth, flowering and fruiting and quality of the pineapple fruit (TSS, acidity, sugars vitamin A and C) under tropical, rainfed and humid conditions of Andaman and Nicobar, India were significantly improved by application of 14 g N, 6 g P and a 12 g g K/plant/crop in var. Kew and 10gN and 3g P and 6 gK/plant/season in Var. Queen, (Singh *et al.*, 2002).

5.9 Nutrient Interactions

It is a well-known fact that a nutrient applied will not only result in increased content of the same in the plant, but will also bring about changes in the contents of other nutrients. This phenomenon is called nutrient interaction. A nutrient can have a favourable or depressing effect on other nutrients. Hence, application of other nutrients have to be monitored in relation to their interactions with nutrients applied. In view of the overlapping and opposite roles of nutrients, their optimal proportions are of more importance than their absolute quantities for successful pineapple production, particularly of the 3 most important cations, *viz.* potassium, calcium and magnesium.

NH_4 ions as well as K ions have an unfavourable influence on absorption of Ca and Mg. Scharrer and Jung (1955 a, b) on the basis of the influence of nutrition over the proportion of cations and anions in the plant concluded that the proportion is always constant, that is, no more cations are absorbed than the equivalent quantity of anions that are absorbed or formed (organic acids) by the plant due to the existence of the phenomenon called base-equilibrium constancy in every plant system. Therefore, higher concentrations of potassium led to a reduction in absorption of Ca and Mg (Sideris and Young, 1945) in pineapple. If nitrogen is supplied in the form of ammonium salt as cation, then proportion of anions to cations in nutrient solution is displaced. In contrast to other cations, ammonium ions decisively influence absorption of anions (Scharrer and Jung, 1955 b). The increased phosphate content (Sideris and Young, 1945) and greater acidity of pineapple-plants nourished with ammonium are explained by this. The tendency for adjustment to constant values of cations: anions is also present in the absorption of ammonium (Scharrer and Jung, 1955 b). With ammonium (NH_4) nutrition, quantities of Ca and Mg absorbed by the plant must, therefore, be less than with nitrate nutrition; as is confirmed by Sideris and Young (1946).

5.10 Yield and Yield Factors

Influence of Cations on Flowering

When acetylene treatment was given for flower induction in pineapple, largest number of plants were observed with cation proportions of 42.5–50 per cent K + 42.5–50 per cent Mg + 0 -15 per cent Ca and the least with 0–25 per cent K + 25–50 per cent Mg and 50 per cent Ca. All combinations of cations in which sum of K + Mg amounted to less than 85 per cent led to a reduction of effectiveness of acetylene treatment. In younger plants, higher Mg and certain Ca content favour flower formation, whereas in older ones, a larger proportion of potash is necessary for flower formation (Martin-Prevel, 1961 a). Py (1959) did not observe any response of potassium and phosphate on effectiveness of flowering.

Influence of K : Mg : Ca on Fruit Weight and Yield

According to Martin-Prevel (1961 a), the average weight of fruit and gross yield per hectare increased with higher applications of K + Mg + Ca. An application of 0.5 g equivalent of cations per plant with proportion of K: Mg: Ca at 42.5 : 42.5 :15 and 50 : 50 : 0 and at constant rates of application of nitrogen and phosphate (5.5 g N and 3 g P_2O_5 per plant for the entire cycle) gave fruits averaging 1.525 kg and over, with a highest gross yield of 43 tonnes/ hectare. However, average fruit weight and yields obtained were not less even with K: Mg: Ca = 15:70:15 and 25: 50:25. These figures show clearly that magnesium plays a very important part in getting higher yields, though significant increases are expected even with dominance of potassium. Too heavy dressings of calcium reduce average weight of fruit as well as yield. The optimum percentage proportion of cations was reported as 42.5 : 42.5 :15 of K : Mg : Ca.

Effect on the Duration of Fruit Development

It has been found that the period between acetylene treatment and ripening of the fruit decreased with increasing application of cations (Martin-Prevel, 1961a). While fruits from unmanured plants could be harvested 190 (fays after acetylene treatment, it took only 175 days with dressings of 0.75 g equivalent of K + Mg + Ca. Alterations in K : Mg : Ca ratio influenced time of ripening. Higher potash enhanced ripening while magnesium and calcium retarded it; and Ca seems to have a greater retarding effect than Mg. The optimum requirements are 70 per cent K + 15 per cent Mg + 135 per cent Ca or 50 per cent K + 50 per cent Mg + 0 per cent Ca.

Effect on the Strength of Fruit-Stalk

A distinct relationship exists between the number of fruits standing erect, and the quantity of cations applied and the proportion of the cations (Martin-Prevel, 1961 a). The higher the proportion of potash and larger the supply of cations, the lesser did the fruit tend to fall over in spite of the higher fruit weight. With higher potash, the diameter of the fruit-stalk increased to a greater extent than the average weight of the fruit. The length of the fruit-stalk also increases with the level of potash application, resulting in no influence on stability. Calcium ions, if their share is more than 15 per cent of the total cations, have an unfavourable effect on properties of stability, diameter and length of the fruit-stalk. Magnesium seems to affect stability of fruit-stalk to a lesser extent than calcium. Stability and diameter of fruit-stalk were almost same at K : Mg : Ca at

70 : 15 : 15, 50 : 50 : 0 or 42.5 : 42.5 :15, but they were less when Mg increased at the expense of K.

Quality Characteristics

Colouration of Fruit

Intensity of colouration of the fruit-skin increases with increasing application of potash (Py *et al.*, 1956). Martin-Prevel (1961 b) confirmed this, besides describing the influence of magnesium and calcium. A higher proportion of magnesium with the same potash resulted in a more intense colouration than a higher proportion of calcium. The occurrence of red and brown pigmentation on fruit-skin is linked with cation nutrition of pineapple. The red pigmentation which imparts desired colour to fruits is more with higher applications of potash. The accumulation of brown pigments around fruitlets, which imparts unattractive appearance to fruit, is more with higher calcium.

Effect on the Core of the Fruit

The core (central cylinder) must have a diameter as small as possible in order to obtain a larger fruit-flesh. With higher application of potash, it is expected to have an increase in the diameter of the core like that of the fruit-stalk, since morphologically core is a continuation of the fruit-stalk. However, it did not occur. On the other hand, Martin-Prevel (1961 b) observed thickening of core when calcium was dominating over magnesium.

Effect on the Texture of Fruit-Flesh

The texture of the fruit-flesh was fibrous in plants which were given only nitrogen and phosphate. Plants which received generous Ga than K and Mg, produced fruits with least fibre content (Martin-Prevel, 1961 b).

Effect on Juice Content

Juice content increases with increasing application of cations; potassium and to a smaller extent magnesium were the effective cations (Martin-Prevel, 1961 b).

Effect on Acidity and Sugars

The acidity of the fruit juice reduced with Ca and Mg, and increased with higher K applications (Marun-Prevel 1961 b) Sugar content of the fruits does not increase significantly with the levels of K dressings, but acidity of the fruit juice certamly increases, resulting in reduction in sugar to acid content (Py *et al.*, 1956, 1957 a; Su, 1958, 1959).

Effect on Aroma and Taste

Fruits derived from plants that had been treated with a fertilizer mixture in which potassium was predominant were superior with respect to aroma and taste. Magnesium also slightly favoured aroma and taste but calcium was found injurious (Martin-Prevel, 1961 b).

In overall analysis, maintenance of an optimum proportion of nutrients, especially cations K: Mg : Ca is a pre-requisite for higher of quality fruits. Though high yield and good quality is difficult to combine with one another, it is possible in pineapple to do so, since both yield and quality of the fruit are promoted by higher proportion of potash in the cation mixture.

Finally, it can be concluded that optimum quality can be obtained with 42.5 per cent K, 42.5 per cent Mg and 15 per cent Ca.

5.11 Nutrient Removal

Pineapple removes very large amounts of soil nutrients. Repeated cultivation on the same plots leads to severe reduction in yield after a few years, as a consequence of the great exhaustion of the nutrient, reserves of the soil. Hence, heavy dressings with fertilizers are inevitable to maintain productivity.

Total quantity of nutrients removed by a pineapple-crop from the soil is obtained from an analysis of the plant dry matter, by determining the nutrient contents of the plant substance analysed and the weight of the plant material produced per unit area (Table 10).

The nutrient uptake depends on the variety grown and is influenced by environmental and cultural factors such as soil, climate, planting density, fertilization and yield.

Relative Quantities of Nutrients

Pineapple has a high demand for nitrogen and potash. However, phosphorus is absorbed by the plant in relatively small amounts.

The proportion of N : P_2O_5 : K was calculated based on removal figures. The proportion on an average amounted to 1:0.40 :3.5. The ratio of nitrogen to potash in nutrient removal by the plant is very wide. The amount of nutrients removed per tonne of fruits harvested was 0.75-0.80 kg nitrogen (N), 0.15 kg phosphorus (P_2O_5); 2.0-2.6 kg potassium (KjO), 0.15-0.20 kg lime (CaO) and 0.13-0.18 kg magnesium (MgO) (Martin-Prevel *et al.*, 1962; Aldrigo, 1966: Black and Page, 1969; Lacoeuilhe, 1976). It shows importance of potash to pineapple for its successful growing.

Table 10: Nutrients Removed by Pineapple (kg/ha)

Authors	N	PA	K_2O	Ca°	Mg°	Remarks
Krauss (1928)	350	121	1,131	245	–	18, 375 plants/ha
Follet-Smith and Bourne (1936)	107	87	417	113	74	25,000 plants/ha
Boname (1920)	83	28	437	–	–	12,500 plants/ha
Cowie (1951)	123	33.5	308	–	–	100 tonnes/ha
Martin–Prevel et al. (1961c)	67.5	24.5	174	27	16.2	38,500 plants/ha with 55 tonnes/ha fruit yield

Table 11: Nutrient Removal Proportions of Pineapple Calculated from Reports of Different Authors

Authors	N	:	P_2O_5	:	K_2O
Krauss (1928)	1	:	0.35	:	3.2
Follet-Smith and Bourne (1936)	1	:	0.81	:	3.9
Boname (1920)	1	:	0.34	:	5.3
Cowie (1951)	1	:	0.27	:	2.5
Martin-Prevel et al. (1961c)	1	:	0.36	:	2.6
Mean	1	:	0.40	:	3.5

Leaf Analysis

Fertilizer is one of the major inputs, accounting towards higher cost of cultivation. With increasing cost of fertilizers, it is imperative to use them more efficiently and judiciously, not only to obtain higher yields per unit of fertilizer inputs, but also to avoid problems of environmental pollution due to their considerable build up in soil and accumulation in drinking water. It is in this context that leaf analysis has proved very useful in monitoring required nutrient status of pineapple at different stages of growth, and this is being widely used in crop-logging technique in Hawaii and Australia. Nutritional status of the plant as revealed by leaf analysis, which is carried out 2-5 months after planting, is the current basis for evolving individual fertilizer programmes for each plantation.

Leaf Sampling

The application of leaf analysis is largely dependent on the type of tissues sampled. Irrespective of the type of sample used *i.e.* the middle-third of the non-chlorophyllous part

(Hawaiian technique) or the whole leaf (IRFA technique-Research Institute for Fruits and Citrus Fruits, Montpellier, France), the 'D' leaf is always used, as it is the only leaf mat can be easily identified and that provides a reliable and sensitive indication of plant nutritional status (Py *et al.*, 1987). The 'D' leaf can also be used to estimate growth which is indispensable for interpretation of leaf analysis. In India, basal white portion (devoid of chlorophyll) of the 'D' leaf is generally employed for depicting nutrient status of pineapple at a given time more meaningfully in relation to yield. The 'D' leaf is defined as the most recently-matured leaf with maximum physiological activity. The vigour and nutrient contents of these leaves have been found to be highly correlated with growth and yield potential of pineapple-crop (Sideris and Krauss, 1939).

The 'D' leaves are sampled during vegetative growth only. Leaves can be sampled once the plant or sucker is large enough, and can be tested until flower emergence. After flowering, plant has only mature leaves which are not suitable for analysis.

Sampling is usually done shortly before foliar sprays are applied, when levels are likely to be the lowest, and previous sprays have had sufficient time to be absorbed. Sampling has to be done at least 2 weeks after fertilizer application.

Leaf Nutrient Standards

In Hawaii, the highest yields and fruits of best quality were obtained when K : P ratio in the base of the 'D' leaf amounted to 12:1 (Cooke, 1949). Higher yields in Red Spanish Pineapple at Puerto Rico were obtained when leaf

N, P and K were 1.66 per cent, 0.14 per cent and 125 per cent (Samuels *et ai*, 1954); and there was no significant yield increases beyond these limits. In Taiwan, Pan (1957) found normal leaf nutrient values to be 1.5-2.0 per cent N, 0.7-0.8 per cent P and 3.5-4.0 per cent K; in Puerto Rico, De Geus (1973) reported optimum leaf nutrient levels (per cent dry matter) for 2 cultivars by analysing basal part of the D-leaf as 1.7-2.2 per cent N, 0.20-022 per cent P and 3.5-4.0 per cent K for Red Spanish cultivar and 1.6-1.9 per cent N, 0.16-020 per cent P and 1.8-2.5 per cent K for Smooth Cayenne. Much lower figures were found in Ghana for cv. Sugarloaf. They were 0.35-0.40 per cent for N; 0.04 per cent for P and 0.44 per cent K (De Geus, 1973). This shows that cultivars do exert some influence with regard to their nutrient requirement. The sufficiency requirement established for Smooth Cayenne from South Africa was 150 per cent -1.70 per cent N; 0.09-0.12 per cent P; 2.20-3.00 per cent K; 0.40-1.00 per cent Ca and 0.30-0.50 per cent Mg (Langenegger and Smith, 1978).

In peninsular India, a close correlation of N and K contents of the middle one-third portion of the D-leaf base of Kew pineapple, sampled at the fifth month of plant growth, was found to exist with fruit yield (Subramanian *et al.*, 1972). It was observed that nitrogen content of 1.4 per cent and potassium content of 3.7 per cent on dry weight basis in the middle one-third portion of the leaf-base sampled in the fifth/Sixth month could account for 90-95 per cent of the yield (Subramanian *et al.*, 1974). This can be utilized in increasing productivity. If N and K are lower man mentioned above, the nutrient status of the plant could be effectively built up for yield increase. In Queen pineapple, nutrient status *A* eleventh month was better correlated with yield than at seventh month. Based on the analyses, the critical levels of nitrogen in the middle one-third of the basal leaf base (base N) sampled at fifth, eighth, and eleventh months of plant growth were found to be 1.5 per cent, 1.23 per cent and 1.97 per cent. Whereas in the remainder of fee D-leaf at fifth, eight and eleventh months, the critical N levels were 0.99 per cent, 0.81 per cent and 1.37 per cent (Hariprakasa Rao *et al.*, 1977). Since it is advisable to take up leaf analysis at an early stage of plant growth, to build up nutrient status of the plant for higher yields through fertilization, sampling at fifth month appears to be more appropriate. Suggested leaf nutrient guides for pineapple from different places are given in Table 12.

5.12 Manuring and Fertilization

Pineapple removes 123 kg of nitrogen, 33 kg of phosphorus and 308 kg of potash from one hectare of land yielding a crop of 40 tonnes. It, therefore, requires abundant supply of nitrogen and potash. Increased yield and improvement in the quality of pineapple by application of fertilizer have been reported from the pineapple growing areas in the different parts of the world.

Krauss (1928), Follett-Smith and Bourne (1936) reported that pineapple plants required higher amounts of nitrogen and potash than phosphorus. Vasconcelos (1952) in Brazil obtained the highest yield with 120 kg N, 60kg P.,0;, arid 120 kg K_2O per hectare. Treatment with 560 kg N, 140 kg $P;O_S$ and 560 kg K_QO per hectare was found adequate for a plant density of 42.000 per hectare (Su, 1957a). He further suggested that the most economic rate of N ranged from 12 to 18gm per plant, and 40 per cent of this was required for ratoon crop. According to Srivastava (1963) nitrogen

Table 12: Suggested Leaf Nutrient Guides for Pineapple from Different Places

Variety	Place	Stage (Months after planting)	N (%)	P (%)	K (%)	Ca (%)	Mg (%)	Authors
Red Spanish	Puerto Rico	–	1.66	0.14	4.25	–	–	Samuels et al. (1954)
	Taiwan	–	1.5–2:0	0.7–0.8	3.5–4.0	–	–	Pan (1957)
Red Spanish	Puerto Rico	–	1.7–2.2	0.20–0.22	3.5–4.0	–	–	DeGeus(1973)
Smooth Cayenne	Puerto Rico	–	1.6–1.9	0.16–0.20	1.8–2.5	–	–	DeGeus(1973)
Smooth Cayenne	South Africa	–	1.5–1.7	0.09–0.12	2.2–3.0	0.4–1.0	0.3–0.5	Langenegger and Smith (1978)
Sugarloaf	Ghana	–	0.35–0.40	0.04	0.44	–	–	DeGeus (1973)
Kew	India	Fifth month	1.4	–	3.7	–	–	Subramanian et al. (1974)
Kew	India	Fifth month	1.51	–	–	–	–	Hariprakasa
		Eighth month	1.23	–	–	–	–	Rao et al. (1976)
		Eleventh month	1.97	–	–	–	–	Rao et al. (1977)

phosphorus and potash should be applied in the ratio of 2:1:1. The results of the experiments conducted at Bidhan Chandra Krishi Viswavidyalaya indicated that nitrogen at the rate of 12-16 gm, phosphorus at the rate of 2-4 gm and potassium at the rate of 10-12 gm per plant should be the optimum for a plant density of 51,000 per hectare (Khatua *et al.*, 1988), while Roy (1981) obtained the highest yield with N. P and K at the rate of 600. 200 and 600 kg per hectare respectively, in plant density of 64,000 and 72,727 per hectare. Foliar application of urea at a concentration of 4 to 5 percent in dry season and 10 percent in wet season may be made.

Marchal (1971) explained that the critical level of leaf phosphorus of the 'D'' leaves was 0.05 per cent for all stages of growth and phosphorus deficiency affected the utilisation of nitrogen. Tay *et al.* (1969) also concluded that high levels of nitrogen increased the nitrogen content in leaf but decreased the potassium content. Khatua *et al.* (1988) obtained the highest yield with 16 gm N, 4gm P and 16 gm K. per plant which indicated a K:P ratio in the leaf at the flowering stage as 20 : 1.

Among the different sources of nitrogen, the effectiveness of ammonium sulphate as compared to urea and calcium ammonium, nitrate in increasing the yield and size was recorded by Chadha *et al.* (1974). Calcium ammonium nitrate was found to delay the maturity of fruits. Treatment with ammonium sulphate showed a reduction in fruit.acidity (Khatua *et al.*, 1988).

Cibes and Samuels (1958), revealed that the formation of yellowish snots on the margins of older leaves was caused due to magnesium deficiency. Magnesium was found to be the limiting factor for yield in sandy and strong acid soils. On the basis of yield and quality of pineapple fruit. Martin-Prevel (1961) determined the optimum proportion of three fertilizer K : N per cent : Ca at 42'5 : 42'5 : 15'0. Su (1956) reported that the critical values of magnesium were 0.22 per cent in leaf tissue and 60 ppm of exchangeable magnesium in the soil. Application of zinc at 2 ppm extended the cropping period and proved fruit quality (Srivastava, 1960). Tay (1974) observed that death of apical region of plant was due to boron deficiency. Soil application of magnesium (164 gm/plant), boron (0.24 gm/plant) and iron (0.64 gm/plant) hastened flowering by 8-9 days and an increased yield was obtained with magnesium (1.28 gm/plant) and boron (0.48 gm/plant) (Sen *et al.*, 1985).

Application of fertilisers in two split doses, once at the onset of monsoon (May-June) and again at the end of rainy season (September-October) after the fruits are harvested and slips and suckers are removed, has been found effective to promote growth and yield. Fertilization is followed by earthening around the stem (Singh, 1998).

5.13 Methods of Fertilizer Application

Primarily there are 3 methods of fertilizer application in pineapple, *viz.* soil application, foliar application and liquid fertilization.

Soil Application

It is most widely followed in India and other countries, where the use of paper or polyethylene mulch is not followed. In Cuba, studies on Red Spanish cultivar with 3 methods of application revealed that application of fertilizers in bands along the rows was economically more viable than applying fertilizers to axils of the lower leaves and foliar spray (Nunez Soto and Garcia Serrano, 1978). Experiments conducted at the Indian Institute of Horticultural Research have shown that application of part or full dose of nitrogen in the form of foliar spray of urea did not prove superior to nitrogen applied to soil alone (Reddy *et al.*, 1983).

Foliar Application

This is a common practice in several countries such as Australia (Cannon, 1960; Mitchell and Nicholson, 1965), China (Su and Huang, 1956), French Guinea (Py, 1962), Hawaii and the Philippines, wherever mulching is a common practice. Leaf spraying is well adapted to pineapple morphology since trichomes play an important role in absorption of elements in solution (Sakai and Sanford, 1980). Experiments with ^{15}N (Marchal and Pinon, 1980) have shown higher absorption of nitrogen on the underside of leaf, which has more trichomes. The younger is the leaf and trichomes, the higher is the rate of absorption. First dose of fertilizer is applied before spreading mulching material and later doses of fertilizers are applied through foliage or liquid fertilization. In Kenya, a basal application of NPK to soil prior to spreading plastic mulch and foliar application of urea from fourth to fourteenth month was found to increase greatly total yield and also Grade I fruits (Waithaka and Puri, 1971). In pineapple, foliar application is of special importance due to spiny leaves and high-density plantations.

Liquid Fertilization

In some plantations in Hawaii, ammonium sulphate and potassium sulphate are applied to axils of lower leaves in the form of a solution; in which a boom-type-sprayer is employed. Fertilizers reach leaf and soil in the form of solution and can be absorbed directly by the plant. Since fertilizers are applied in a considerable diluted form to avoid scorching of younger leaves, it also provides considerable quantities of water at the disposal of plants during dry seasons. To sum up, soil application was found the best method of manuring pineapple. However, in places where use of polyethylene or plastic mulches is prevalent, supplementary manuring of nitrogen through foliar application of urea seems to be relevant.

Chapter 6
Interculture

6.1 Earthing Up

This is an essential operation in pineapple cultivation aimed at good anchorage to plants. It involves pushing soil into the trench from the ridge, where trench-planting is a common practice. As pineapple roots are shallow, plants lodge in heavy rainfall, especially in flat-bed plantations. Lodging of plants when fruits are developing on them would result in lopsided growth, uneven development and ripening of fruits. This operation becomes more important in ratoon crops as the base of the plants shifts up crop after crop (IIHR, Bangalore, 1977). Close planting with 63,000 plants/ha would minimize lodging as plants prop one another (Chadha *et al.*, 1974).

6.2 Weed Control

Successful weed control is very important in pineapple growing. Hand-weeding is common in pineapple culture in India. In tropical areas, weeds grow profusely and hamper growth of pineapple-plants, especially during early stages of plant growth. Therefore, quite a number of weedings are necessary in a year to keep pinery free from weeds. The major factor which contributes to high cost of production of pineapple is the manual weeding; which accounts for nearly 40 per cent of the total cost of production. A total of 130 tonnes/ha of weeds were collected in 3 manual weeding operations over the course of a year, during an experiment carried out in Cameroon (Gaillard, 1971). Pinon (1976) obtained mean yield of 14 tonnes/ha in unweeded plots, as compared to 79 tonnes/ha in manually weeded plots, and as opposed to 83 tonnes/ha in plots treated with chemical herbicide and with supplementary hand-weeding. The non-availability of labour and high wages during recent years have aggravated problem of pineapple cultivation and its extension to new areas. Hand and mechanical weeding were used primarily until oil emulsion sprays were developed in 1940.

Soil-mulching also has been tried as means to check weed growth. In Hawaii, early attempts to control weed with paper, straw and polythene-film mulches were successful (Savage and Barnett, 1934; Collins, 1960). Besides using mulches for weed control, the field layout has to be specific, and special machinery is required for this purpose.

The control of weeds with herbicides offers a good alternative to manual weeding, especially in pineapple-plants, as leaves are spiny, and may restrict movement of labourers under high-density plantations. Weeds must be controlled in pineapple by shallow weeding or by sprays, as deep cultivation would interfere with shallow-root system. Thus, chemical weed control has an added advantage here.

The first attempts at chemical weed control with 2, 4-D were unsuccessful as it was found to affect flowering adversely (Crafts and Emanuelli, 1948). During fifties, pre-emergence herbicides such as Manuron (CMU) were developed; and then selective materials for the control of weeds, those escape pre-emergence sprays, were. The next development in chemical weed control was Pentachlorophenate alone or in diesel oil emulsion, as early as 1951 in Queensland. Pentachlorophenol (PCP) (1.36 kg) and sodium pentachlorophenate and one gallon of mineral oil or creosote emulsion/100 gallons controlled broad-leaved weeds, but higher concentrations were needed for grasses (Cannon, 1952). The first good pre-emergence chemical for weed control was Manuron (CMU). Good control of weeds was obtained with 27-55 kg/ ha of CMU as a pre-emergence ground spray (Wolf, 1953). Py (1955) obtained good results on Asteraceae with 2.5 kg (80 per cent active) CMU/hectare. Cannon and Prodonoff (1959) found 4 Ibs of CMU/acre to be better than 20 Ib/acre of PCP. No injury to pineapple occurred with even up to 8 Ib/acre of CMU, but 8 Ibs gave no better control than 4 Ibs. Diuron is an improvement over CMU. It is safe in rainy season; CMU can be used only during dry season (Barbier and Trupin, 1956). Diuron at 2.5-5.0 kg/ha seemed to be successful in controlling weeds (Py, 1959 b), provided grasses were destroyed before planting either mechanically or with Dalapon. Cannon (1960 b, 1960 c) found Diuron and Atrazine promising and superior to CMU, PCP and Simazine, whereas TCA and Dalapon were injurious even at 1.6 kg/ha. Effective weed-control can be achieved by the use of more than one chemical. Guyot and Oliver (1958) found best results with 6 litres of PCP and 3 kg of Diuron in 52 litres of orchard spray-oil. In the Philippines, Manuel (1962) reported that CMU and Simazine at 2, 4 and 6 Ibs in 30 gallons of water/acre gave satisfactory control in very dry areas. In Ivory Coast, Simazine plots (5 kg/acre) had the least number of weeds, 20 weeks after treatment. Diuron (80 per cent active) at 25 kg/ha was almost as effective and even superior to CMU but PCP was not at all effective

In New South Wales, paper mulch was found beneficial in very dry seasons for maintaining higher top-soil moisture and allowing plants to feed on the top soil (Savage and Barnett, 1934). With this, flowering and fruit maturity occurred 2-3 weeks earlier, and yields and fruit quality were better than control, and weed problem was very much reduced. Magistand *et al.* (1935) compared a mulch of bagasse (crushed sugarcane stalks) with paper mulch, the one commonly used. The yield and soil moisture were highest under bagasse mulch, and paper mulch improved nitrogen content of the soil and kept the soil warm, particularly in winters. Warm temperature in root zone is very favourable for pineapple. Covering ground with tarred paper raises soil temperature and prevents fields being infested with weeds, especially during the first twelve months after planting, before the leaves of the pineapple-plants shade ground sufficiently, besides preventing losses of N by leaching and water evaporation. The yields increased by 22-34 tonnes/ha with paper mulch. Plant nutrients, however, are not supplied by paper mulch, as is done by bagasse and other organic mulching materials. In Taiwan, Su *et al.* (1956-57) tested effect of rice-straw mulch on the soil, yield of the pineapple and quality of fruits. Mulching increased moisture content of the soil, amount of available potassium and nitrogen and the organic-matter content. N, P_2O_5 and particularly K in the leaves increased with increased applications of straw. If soil was mulched with rice-straw only between the rows, the yield stood at 33 per cent above that of the control. All over mulching raised yield by 50 per cent. The production of fresh sound fruits with firm tissues and high sugar content increased by 133 per cent. In regions with a low rainfall in Hawaii, mulching with pineapple trash has been employed in combination with a paper mulch (Collins, 1960). The trash is produced by chopping and crushing old pineapple-plants which remain on the field at the end of the crop cycle, and placed over the soil surface, without the use of paper; or it may be placed between the papered beds, or both, on and between the papered beds. The most commonly used method combines paper mulch on the beds and trash in spaces between beds. Trash mulch should not be used in regions with adequate rainfall, as it acts as a mediator in the spread of fungal diseases. Use light weight paper mulch is commonly followed in Australia (Cann, 1961). Black polythene was found significantly superior to straw as a soil covering, for limiting water loss in dry sea in Guinea (Py, 1965). Straw depressed vegetative growth, when polythene favoured, if not significantly, growth and yield.

Mulching is not a common practice in India. Use of dry leave; straw is in practice in some areas of south India (Naik, 1963). Preliminary trials conducted at the Indian Institute of Horticultural Research Bangalore, revealed that mulching with black polythene and saw dust resulted in better growth of plants as compared to white polythene ;

paddy-straw mulching (Sharma *et al.*, 1971). Chadha *et al.* (1974) recommended high-density planting to minimize need of mulching.

Benefits accrued through paper mulch are as follows:

1. It helps to keep soil friable and warm during cold weather;

2. It reduces weeding costs since weeds cannot grow in darkr under paper;

3. It conserves moisture especially of the surface layers of the where the roots of the young plants are developing. Conserved moist ensures constant availability of nutrients present in the soil;

4. It reduces leaching of soluble N and other nutrients;

5. It results in vigorous and uniform growth of larger and heart plants, earlier crop and better propagules;

6. It ensures better quality fruit and higher yields (Collins, 1960, Cann, 1961).

Though paper mulch has been used with success for a number of years, it has now been largely replaced by black polythene film of 0 0.05 mm thickness, which lasts longer and has the same advantage (Samson, 1980).

6.3 Removal of Suckers, Slips and Crowns

Suckers start growing with the emergence of inflorescence and grow with developing fruits. As one or two suckers, depending on plant density, need only to be retained on the plant for ratoon additional suckers and all slips are removed; as the growth of these is weaken plant and hinder fruit development. Fruit weight increased with increasing number of suckers per plant, while increase number of slips delayed fruit maturity. Crown size has no bearing fruit weight or quality (Chadha *et al.*, 1977). Hence, desuckering can be delayed as much as possible, while slips are recommended for removal as soon as they attain the size required for planting. In a situation where early harvest is required or slips are not required for planting, they can be removed as and when they sprout. Removal of crown is not required as it mars appeal of fruit and also makes handling difficult. However, reduction of crown size is necessary in Smooth Cayenne, if the fruits are intended for export. The meristem of the crown is gouged out with a chisel when it is about 8 cm long to restrict over sizing of crowns. One man can perform this operation on 1,000 to 2,000 plants per day (cited by Samson, 1980). Partial pinching of crown in Kew pineapple consisting of removal of the innermost whorl of leaflets along with the growing tips 1½ months after fruit set was found best to get better fruit size and shape (Prakash *et al.*, 1983).

Chapter 7
Irrigation

7.1 Water Requirement

Pineapple is grown mostly as a rainfed crop in heavy rainfall areas. Optimum range of rain needed for pineapple is 1,000 to 1,500 mm. However, some of the pineapple-growing areas come under high-rainfall zone, where rainfall is to an extent of 3,000 mm (in Cameroon and Costa Rica), between 2,600 and 2,900 mm (in Malaysia) and nearing 2,000 mm (in Taiwan). As pineapple roots are very sensitive to water logging, good drainage is necessary. Potential evaporation figures range from 1,472 mm in Malaysia to 839 mm in South Africa (Bartholomew and Kadzimin, 1977). A potential evapotranspiration of 3.5-4.5 mm/day was reported for Vamoussokro at 7°N latitude in Ivory Coast.

As such pineapple is a xerophyte, having many adaptations to drought. Its stomata close during day when much moisture can otherwise be lost. Its leaves have stiff and waxy upper surface and have stomata in furrows on the lower surface; protected by thick growth of hairs called trichomes, which minimize loss of water through transpiration. Leaves are also arranged in a rosette, very close to each other, on a condensed stem in such a way that rainwater collected gets conducted to the base and will hardly pass to the ground. Similarly, dew collected at the leaf bases is also absorbed. Thus, pineapple utilizes thoroughly the available water in the atmosphere and needs very little water through irrigation. Hence, Rao *et al.* (1977) termed pineapple 'A boon for dryland agriculture'. As water-holding capacity of the soil rarely surpasses 100 mm and potential evapotranspiration from the plant can go up to 4.5 mm per day, pineapple exhausts water supply in 3-4 rainless weeks (Samson, 1980). Under such conditions, even though the plant is drought resistant, but the dry conditions can lead to delayed growth and fruiting. Hence, supplementary irrigation in drier periods and in low rainfall areas is beneficial. As such irrigation is supposed to be a luxury for pineapple. A total of 14-22 tonnes of pineapples per hectare were harvested additionally with supplementary irrigations in Ivory Coast (Samson, 1980). According to Sideris and Young (1945), pineapple-plants with an average weight of 4,287 g require 60,860 g of water from the commencement of growth to the time of blossom formation for transpiration and development of plant substance alone having 42,500 plants/ha with 4.45 cm rainwater per square cm. In this, losses of water from soil by seepage and surface run-off or evaporation are not taken into account. In some places in Hawaii, where rainfall exceeds 2,540 mm (100 inches), it affects quality and keeping properties of fruit (Malan, 1954). In Hawaii, a rainfall of 1,100-1,300 mm (45-55 inches) is regarded as optimum (Cooke, 1949). Pineapple needs a well-distributed rainfall of about 1,200-1,500 mm annually (Giacomelli, 1967). Application of 120 mm water monthly in 4 doses has not been found superior to 60 mm, but both treatments significantly improved growth, flowering, yield and sucker production. Although pineapple is grown in India in rainfed areas, where sufficient rainfall is received, but it can also be grown successfully with a few irrigations during summer in the semi-arid tropic regions of the country, where average rainfall varies between 1,000 and 1,200 mm annually, distributed fairly in the rainy season, as around Bangalore. A scientific approach towards finding out water requirement of pineapple was made by Rao *et al.* (1974) at the Indian Institute of Horticultural Research. As a result, it was observed that 80 per cent depletion in the available soil moisture is as good as 20 per cent depletion in the available soil moisture or in other words an available soil moisture regime of 100–20 per cent was sufficient for Kew pineapple. Therefore, in scanty rainfall areas and years and during hot weather, irrigating pineapple once in 20-25 days is advisable wherever facilities are existing to ensure a good crop. Experiments conducted at Chethalli (Coorg) revealed that 4 irrigations during summer were desirable, which not only increased yield by 20 tonnes/ha but also advanced plant-crop cycle by 7-9 months by way of increased plant growth and advanced physiological maturity (Singh *et al.*, 1977). Irrigation can also be helpful to establish off-season planting to maintain year-round production of fruits for feeding canning factories.

7.2 Methods of Irrigation

There are 5 methods of irrigation in vogue. These are flat-bed irrigation, furrow irrigation, trench irrigation, sprinkler irrigation and drip irrigation.

Flat-bed Irrigation

This is followed in areas where pineapple is planted in flat-beds and soil surface is even. Uneven distribution and wastage of water and concentrations of applied nutrients at one side of the field are drawbacks of this system.

Furrow Irrigation

This system is practised in places where pineapple is planted in furrows and topography is of gentle slope. Water is let in from one end of the furrow to the other. Accumulation of applied nutrients at the farther end of the furrow is the major problem.

Trench Irrigation

This is similar to furrow irrigation except that water is let in between rows of the crop in a trench.

Sprinkler Irrigation

This system is not very common in India, excepting in commercial pineries owned by canning industries. Though this involves high initial investment, it ensures efficient use of water. Moreover, it can be employed to irrigate in highly undulated terrain of land where conventional systems are not successful. Trials carried out elsewhere have revealed that sprinkler irrigation of pineapple fields in Taiwan during dry season at intervals of 2, 4 or 5 weeks significantly improved plant growth and weight of spring- and summer-maturing fruits in proportion to the frequency of irrigation

(Huang and Lee, 1969). Low volume but frequent sprinkling of water by this system increases humidity of microclimate around plants which is highly desirable. The added advantage with sprinkler irrigation is that application of fertilizers and insecticides can be carried out along with irrigation.

Drip Irrigation

It requires high initial investment. There is one pipe per row of plants, *i.e.*, approximately 8,000 metres of pipe is needed to irrigate 55,000 to 60,000 plants/ha (Py *et al.*, 1987). Since it is expensive, cost-effectiveness of this system depends on the regular inclusion of fertilizer and pesticide in the irrigation water and production of two or even more successive harvests. This system also has drawbacks as water has to be filtered and the flow needs to be controlled, depending on the slope of the ground.

Chapter 8
Inflorescence, Fruit Development and Yield

The reproductive phase of the pineapple begins in response to natural or plant-growth-regulator-forced induction of reproductive development (natural induction and forcing). Because the inflorescence of pineapple is terminal, when reproductive development begins, formation of new leaves ceases. Expansion of previously initiated leaves continues, but not all of these expand fully. Some of the leaves are found on the fruit peduncle and their size is much reduced relative to older fully expanded leaves. Once reproductive development is initiated, inflorescence and fruit development continue without interruption until the fruit matures, although development may be interrupted by disease or slowed by water or temperature stress. Fruit growth ceases at air temperatures below about 10°C and very high air temperatures–certainly above 35°C–can also retard growth (Malezieux *et al.*, 1994).

The meristematic area of the pineapple stem is small relative to the diameter of the stem near its apex. The vegetative apical dome is very small, but it broadens considerably after inflorescence induction has occurred. At the onset of reproductive development, leaf primordia width decreases and the number of primordia bordering the dome increases considerably (Kerns *et al.*, 1936; Bartholomew, 1977). At this early stage, the first structures of the inflorescence to develop are bracts/and each fruitlet forms in the axil of these bracts. Shortly after the bracts have been initiated, three sepal and petal primordia can be seen' developing and, soon-thereafter, six stamen primordia develop. The elongation of the stem tip in a cross-section taken through the centre of the plant is diagnostic of natural or growth-regulator-induced inflorescence development and can usually be seen within 2 weeks after induction has occurred. Development of the florets at the base of the young inflorescence continues at the same time that new bract and flower primordia are being produced higher on the young inflorescence. Eventually all flower parts become enclosed by the developing sepal primordia and are no longer visible without dissection. The number of florets that develop on a fruit varies considerably with the variety of pineapple, the size of the plant at induction, plant population density, the quality of forcing and other factors that have not been adequately characterized. For 'Smooth Cayenne' pineapple, at between 30 and 40 days after induction, the diameter of the apical dome decreases, the initiation of reproductive structures ceases and leaf bracts and, later, crown leaves begin to develop (Kerns *et al.*, 1936; Bartholomew, 1977). By the time crown leaves are visible, there has been considerable elongation of the peduncle and the leaves (bracts) in the centre of the leaf whorl have turned red to reel-orange. The rate of early development of the 'Smooth Cayenne' inflorescence and fruit is determined almost entirely by the prevailing temperature of the environment where the crop is grown (Fleisch and Bartholomew, 1987; Malezieux *et al.*, 1994). Cell division continues in the fruit until after anthesis and further development thereafter is primarily the result of cell enlargement (Okimoto, 1948).

The stages of fruit development after induction that can be defined on the basis of developmental morphology include the beginning and end of floret (fruitlet) initiation, the beginning and end of anthesis (flowering) and maturation. Other 'stages' have been defined for convenience in estimating progress towards maturation or to help in determining when disease organisms might enter the fruit (Rohrbach and Taniguchi, 1984). The earliest stage that can be seen in the centre of the leaf whorl as the peduncle elongates without sacrificing the plant is commonly referred to as 'open heart'and the width of the opening is commonly estimated for crop-logging purposes. At the time the 'open heart' stage is reached, inflorescence bracts have turned bright red to red-orange.

Fruit Growth and Ripening

The time from induction to maturity is a characteristic of the clone or variety and is greater for 'Smooth Cayenne' than for the 'Spanish' and 'Queen' varieties (Py *et al.*, 1987). Earliness was associated primarily with the time required to reach a plant weight suitable for induction. Growth in fruit weight, and presumably volume, is sigmoid (Sideris and Krauss, 1938), with the main biochemical changes associated with maturation, such as accumulation of sugars and carotenoids, occurring in the last weeks of development (Gortner *et al.*, 1967;). With 'Smooth Cayenne', the external signs of approaching maturity are the flattening and increased gloss of the surface of individual fruitlets. As with anthesis, fruitlet maturation occurs progressively from the base to the top of the fruit. As ripening occurs in 'Smooth Cayenne', the fruitlet epidermis typically turns yellow progressively from the base to the top, though, in some locations or at some times of the year, the shell does not colour as the fruit matures and green-shell ripe fruit are produced. Large fruit weighing 2.0 kg or more will have basal fruitlets that are overmature by the time the upper fruitlets ripen so overall fruit quality is near optimum when large fruits are about half yellow. The fruit does not show a respiratory climacteric during ripening, perhaps because it is an aggregate of many fruitlets, all at different stages of development.

Factors Affecting Induction of Inflorescence Development

The success of forcing is related to plant sensitivity to induction, *i.e.* the probability of the occurrence of natural induction. The higher the plant sensitivity, the more likely that forcing will be successful. Success is measured by the fact of induction (percentage of plants induced), the quality of the induction (fruitlets initiated per unit of plant weight) and the sharpness of the harvesting peak. Fruitlet numbers per fruit is established at induction and, thereafter, fruit growth and enlargement become the determinants of yield. Fruit growth and enlargement are optimum when conditions for plant growth are optimum. Maturation of fruits within a field has a natural spread over time and, where forcing is poor, fruit mature over a wider time span than where forcing is good. The sharpness of the harvesting peak affects the number of fruits actually recovered from the field and the cost of harvesting.

Natural Induction of Flowering

'Smooth Cayenne' pineapple must attain some minimum plant weight before natural induction of inflorescence development can occur (Py *el al.*, 1987). The minimum weight has not been well characterized but probably exceeds 1.0 kg in most environments and is greater in warm than in cool environments. Collins (1960) considered 'Smooth Cayenne' to be a perennial that would continue to grow as long as environmental conditions were adequate. There are, however, few regions where environmental conditions are constantly optimum for growth. Seasonal patterns in natural flower initiation are, therefore, evident in most regions. These patterns have long been recognized and were the basis for commercial pineapple production before the discovery of chemical flowering agents.

Once the minimum plant weight is reached, the factors that promote flowering are mainly those that retard vegetative growth. For 'Smooth Cayenne' grown in subtropical regions, there is a strong wave of natural induction in the autumn-winter season. Natural flower initiation can, however, occur at other times of the year (Millar-Watt, 1981; Sinclair, 1997, 1998, 1999), so those plants that do not flower naturally in the winter may flower in midsummer.

Definite peaks in natural initiation are also recognized in tropical regions (Aubert, 1977; Wee, 1978; Giacomelli *et al.*, 1984). The main peak usually occurs when either minimum or maximum temperatures are at their lowest. This peak is also associated with a decline in the hours of solar radiation. A small peak also occurs when maximum temperatures are at their highest (Teisson, 1972). Inhibition of natural induction is associated with those factors that promote vegetative growth. For example, vigorous growth, stimulated by excessive nitrogen and warm night temperatures, inhibits flowering (Nightingale, 1942; Bartholomew and Kadzimin, 1977; Conway, 1977).

Photoperiod

The earliest work on photoperiod by Van Overbeek (1946) showed that pineapple responded like a short-day plant. 'Smooth Cayenne' was a quantitative short-day plant, *i.e.* flowering can occur at any day length but is accelerated by short days (Taiz and Zeiger, 1991). Gowing (1961) observed that flowering of 'Smooth Cayenne' was not inhibited by day lengths in excess of 13 h. He also found that interruption of the dark period suppressed flower initiation, so it was concluded that long nights favoured induction.

Solar Radiation

There is little evidence to indicate that solar radiation *per se* has any direct role in natural flower induction, It seems most likely that natural induction in regions away from the equator is explained by a combination of short photoperiods and cool temperatures. However, natural induction also occurs in equatorial regions, where photoperiod is almost constant and temperatures are generally higher. In equatorial regions such as Cote dTvoire, West Africa (latitude 5-10°N), and tropical western Malaysia (latitude 1°17′N), the periods of greatest natural flowering coincided with decreases in maximum (Teisson, 1972) or minimum temperature (Wee, 1978).

Temperature

While 'Smooth Cayenne' pineapple is considered a short-day plant, cool temperatures in particular enhance the flowering response. It is worth remembering that the optimum temperatures for growth of pineapple are considered to be close to 30°C day and 20°C night with an optimum mean of about 24°C (Neild and Boshell, 1976; Bartholomew and Malezieux, 1994).Van Overbeek and Cruzado (1948b) established that a night temperature of approximately 16°C over 30 days induced flowering in 88 per cent of 'Red Spanish' plants during late summer in Puerto Rico; 28 per cent of plants held at a night minimum temperature of 22°C also flowered. A night minimum temperature of 16"C was therefore more inducive than 22°C for 'Red Spanish'; the optimum temperature for flowering is not known. Gowing (1961), using 'Smooth Cayenne', compared night temperatures of 15, 23 and 26°C and found that a night temperature of 15°C induced flowering when in combination with a short day length for a period of 30 days. At normal day lengths (*c.* 12.5 h) this same low night temperature over 30 days did not induce flowering. However, exposure to a constant 18°C for 9 weeks was claimed to be more effective. The longer period of exposure was probably the main factor accounting for the improved response. A more recent study by Sanewski *et at.* (1998) supports this observation. Sanewski *et al.* (1998) found that a constant 20°C for 10 weeks induced flowering in 100 per cent of 'Smooth Cayenne' plants but a constant 10 or 15°C for up to 12 weeks did not induce flowering. Friend and Lydon (1979) also found that, with a photoperiod of 8 h and day temperatures of 25 or 30°C, a night temperature of 20°C was more inductive for 'Smooth Cayenne' than night temperatures of 15, 25 or 30°C Flowering was induced at night temperatures of 15 and 25°C but not at 30°C. In a review of natural flower initiation of 'Queen' pineapple in Papua New Guinea, Bourke (1976) concluded that natural induction was closely associated with minimum temperatures of 20-23°C. It appears, therefore, that exposure to temperatures of as high as 25°C for at least the nocturnal period, in combination with a short day length will induce flowering in 'Smooth Cayenne' if these plants are exposed to these conditions for 9-10 weeks. A night temperature of 20°C in combination with a day temperature of 20-25°C is believed to be close to optimal.

In addition to the effect of cool temperature, diurnal warming during winter may also increase the incidence of natural induction. In a comparison of six planting densities in subtropical Queensland, Scott (1992) found that the incidence of natural flowering of the ratoon crop increased consistently from 3.5 to almost 12 per cent as plants/ha decreased from 80,695 to 46,112. No leaf- or plant-weight data were taken for the ratoon crop, so it is not possible to rule out differences in plant weight as a factor in this study. Further evidence of an effect of diurnal warming on flower induction is the common observation in subtropical Queensland that natural flowering is more prevalent in outside rows, where incident radiation and diurnal plant temperatures are expected to be greater. There is, however, the possibility that the increased natural initiation is a direct response to increased light interception and not the subsequent rise in plant temperature. While this is possible, data collected by Sanewski *et al.* (1998) suggest that warming after exposure to cool temperatures has a direct positive effect on natural initiation. In their study they collected mature suckers of 'Smooth Cayenne' from the field in late winter and placed them at either 5, 10 or 28°C for 3 days. Exposure of plants to 28°C induced flowering in 75 per cent of plants. There is clearly a need for further study of the effects of temperature and its interaction with solar radiation on plant sensitivity to natural and forced induction.

In addition to the effects of low temperature, it is also commonly recognized that natural induction will occur after brief periods of high temperature. In Australia, these events have been associated with diurnal maxima of 40°C over 2 days (Sinclair, 1997) and is probably the result of wound- induced ethylene production. There is good circumstantial evidence that ethylene is involved in the flower-induction process. Environmentally induced flowering is assumed to result from increased production of ethylene or heightened plant sensitivity to ethylene, or both.

Water Availability

Water stress and water excess are often implicated in a flowering response in tropical and subtropical species. An increase in ethylene production is often associated with water stress in plants (Yang and Hoffman, 1984), but there are conflicting reports on this point. For example, Morgan *et al.* (1990) found only slight, transient ethylene production on rewatering cotton subjected to water deficit but no response in beans or roses. Nevertheless, it is generally believed that pineapple flowering may be induced in response to a seasonal water deficit and induction also occurs in response to water excess (Py, 1964), due to growth inhibition or enhanced ethylene production, or possibly both. Tay (1974) reported that both water excess and water deficit delayed natural induction of pineapple grown in peat soil, while growth and natural induction were enhanced by increasing watering frequency. Consistent with the results of Tay (1974), Min (1995) found that a severe water deficit decreased the plant's susceptibility to ethephon and did not induce natural flower initiation. Mild water deficit was not studied. Water excess increased ethylene production of the basal white tissue of 'D' leaves by approximately 100 per cent but did not induce flowering. The fact that excess water increased ethylene production so substantially suggested it should increase the plant's susceptibility to flowering and may therefore be implicated in some situations. Waterlogging, a more severe degree of water excess, had no effect on ethylene production or flowering (Min, 1995).

Much of the inconsistency in the effect of plant water status on flower initiation is probably due to differences in the severity of deficit or excess imposed. Severe and sudden deficit or excess are likely to result in a cessation of all enzymatic processes, including those involved in flower initiation.

Genetic Variation

Most pineapple genotypes appear to be more susceptible to natural flower initiation than' 'Smooth

Cayenne'. The natural flowering cycle in wild types, such as *Ananas ananas-soides* and *Ananas parguazensis*, is usually annual (Leal and Coppens d'Eeckenbrugge, 1996). In an *Ananas* spp. germplasm collection held in subtropical Australia (latitude 27°S), plants of 'Pernambuco' and 'Mordilona Queen' and the species *A. ananassoides* all flower naturally more frequently than does 'Smooth Cayenne'. Williams (1987) reported that some hybrids flower when they reach a plant weight of about 1.5 kg, regardless of the time of year.

It is assumed that the relatively low incidence of natural flowering in 'Smooth Cayenne' and the associated case of crop control in comparison with other types of pineapple account to a substantial degree for its popularity and may be the most important factor allowing large-scale production of this variety.

Geotropic Stimulation

Van Overbeek and Cruzado (1948a) reported that plants of 'Cabezona' pineapple could be geotropically stimulated to flower, as indicated by a strong flowering response when plants were held horizontally for 3 days or more. It was theorized that flowering was induced by the accumulation of endogenous auxin in the lower longitudinal half of the stem apex, though Salisbury and Ross (1992) note that many gravitropic responses can be attributed to increased sensitivity to auxin rather than an increase in concentration.

In summary, both environmental and cultural factors can contribute to natural induction in 'Smooth Cayenne' pineapple. 'Smooth Cayenne' shows a clear response to day length and temperature and probably a weaker response to solar radiation. Surprisingly, very low temperatures are not commonly implicated in the incidence of natural induction. Flowering is more likely to occur as a result of moderate (around 20°C) temperatures, not much lower than the optimum for growth, sustained over many weeks in combination with a short day length. Ethylene production resulting from mild to moderate wounding of plant tissues due to sudden exposure to short periods of very high or very low temperatures appears a minor cause. Plant water status is probably also implicated in a minor way, though supporting data are lacking.

Forced Induction of Flowering

One of the major impediments in successful cultivation of pineapple is its erratic flowering behaviour. Even after 15-18 months of growth under ideal management, less than 40-50 per cent of the plants normally flower, leading to overlapping of operations and irregular supply of fruits to canning factories. Therefore, it is of utmost importance to regulate flowering for better returns as well as to have a regular supply to canneries. Regulation of flowering will also be beneficial in economizing labour requirement.

Pineapple, being native to tropics, shows almost no periodicity (Collins, 1960). The inference that all tropical plants are insensitive to day-length is incorrect (Samson, 1980). Pineapple variety Smooth Cayenne is a quantitative short-day plant (Gowing, 1961; Py and Tisseau, 1965), and for its promotion of flowering it does not require either diurnal variation in temperature or temperature below 25°C (Friend and Lydon, 1979). Other cultivars react only weakly to day-length but may flower when temperature falls (Bourke, 1976). In Red Spanish cultivar, low temperatures were more effective in promoting floral initiation than reduced day-length (Van Overbeek and Cruzado, 1948). As low temperature usually coincides with short days, these effects are generally confused. When Cayenne plants are sufficiently big, floral initiation takes place closer to short days, and flowers appear 60 days later. Pineapple variety Kew, grown extensively in India, is akin to Smooth Cayenne. Natural flowering in Kew extends from December to February and harvest of fruits from May to July during rainy season. As a result, processing factories experience short supply of fruits during the rest of the year.

Induction of Flowering through Chemicals and Plant Growth Regulators

Ethylene, Acetelene and Calcium Carbide

The effects of these three agents is combined here because pineapple responds to acetylene as it does to ethylene, and calcium carbide releases acetylene when it comes into contact with water. Ethylene properly applied with a pressurized sprayer late in the evening or at night to permit uptake through the stomata when they are open in a highly effective forcing agent, though comparative studies show little difference between ethylene and acetylene. In Queensland, ethylene was used as a saturated solution in 6500-9000 1 ha^{-1} of water or two applications of 4500 l ha^{-1} were applied 24 h apart. Py *et al.* (1987) state that 800 g of ethylene is applied in 6000-8000 of water ha^{-1}. Activated charcoal at 20 g l^{-1} was added to the water to increase absorption of the ethylene in the solution. Plants forced with ethylene produce fruits with physical dimensions similar to those produced by natural induction. Special equipment is required to prepare the solution for application, as the gas *is* highly combustible.

The physiological response to acetylene mimics that of ethylene. Acetylene is more commonly used on small farms because a saturated solution can be made in a pressurized hand or backpack sprayer, using cold or iced water and calcium carbide. Acetylene (Gowing and Leeper, 1959; Yow, 1972) and ethylene (Py *et al*, 1987) were uniformly ; effective in forcing pineapple when applied at night but unreliable when applied during the day, except on cool, cloudy days (Yow, 1972) or late in the evening. A comparative study showed acetylene to be as effective as ethylene, while being cheaper and more convenient to use (Lewcock, 1937). Usually 10-50 ml of solution is applied to each plant (Py, 1953; Py *et al.*, 1987) in the late evening or at night with a pressurized sprayer.

Calcium carbide, which produces acetylene on contact with water, is a highly effective forcing agent, though apparently not as effective as acetylene. The effectiveness of calcium carbide, relative to acetylene, is probably related to the timing and speed of release of acetylene. Consistent with the assumption that acetylene and ethylene move into the plant through the stomata, calcium carbide was more effective if applied at night rather than during the day (Aldrich and Nakasone, 1975). Approximately 1.0g of calcium carbide is applied by hand into the centre of each plant (Py *et al.*, 1987). If no water is present in the centre of the plant, water may be poured into the plant to release acetylene. Hand application limits the use of calcium carbide to small farms.

In pineapple induction of flowering can be achieved successfully which is largely determined by the flowering and fruiting stage of the plant at which plant growth

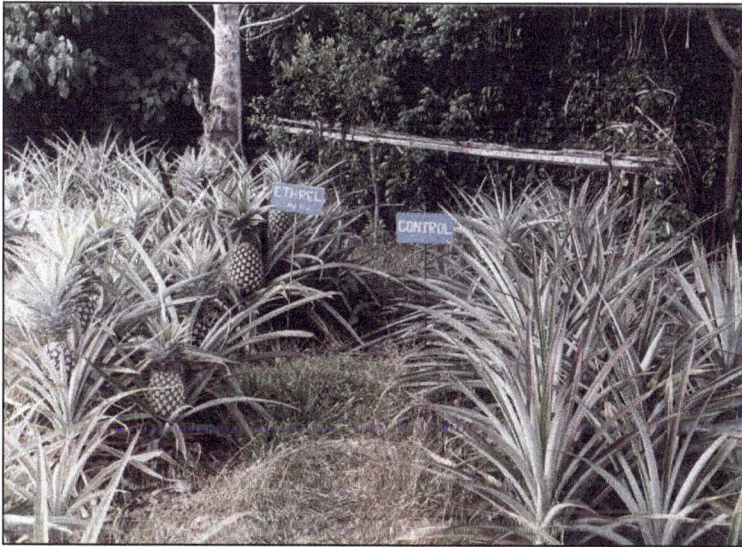

Figure 14: Induction of Floweing by Application of Ethrel

regulators are applied. The treatment of plant growth regulators (Ethrel @ 100ppm) at 40 leaf stage took least days for flowering (43) and harvesting (106) and also produced fruits of maximum weight (2.259 kg) compared to control (0.944kg). TSS, acidity, sugars ascorbic acid and carotenoid contents were improved significantly with increase in number of leaves at the time of induction (Singh and Attri 1999).

Napthelene Acetic Acid

Prior to the discovery of ethephon, the sodium salt of NAA, commonly referred to as SNA in Hawaii, was widely used to force pineapple into flower. Typically, about 50 g of NAA was applied in 900-10001 ha^{-1} of water. While NAA is easy and inexpensive to use, it is less effective than ethylene, acetylene or ethephon, particularly in warm tropical environments. In regions having high average temperatures, NAA may only be effective up to 2 months prior to the time when natural initiation would be expected (Py *et al.*, *1987*) and earlier application may be ineffective. Under less than ideal conditions, double applications spaced about a week apart are essential to force a high percentage of plants. Relative to plants forced with acetylene, ethylene, BOH and ethephon, maturation of fruit of plants forced with NAA is delayed by 1-2 weeks (Yamane and Ito, 1969). It has been speculated that NAA forcing is due to the production of ethylene (Burg and Burg, 1966) and peak ethylene production by plants treated with NAA occurs a few to several days after the treatment. For the above reasons and because plants forced with NAA produce a slightly smaller, more elongated fruit with a somewhat tapered top (Leeper, 1965; Yamane and Ito, 1969), a less than ideal shape for canning, ethephon has largely replaced NAA as the forcing chemical of choice.

The main season of pineapple production in Andamans, India (Tropical and humid) is April to June, but its demand in the market for table use and processing during off seasons, which can be regulated by application of plant growth regulators. Highest percentage of flowering (77 per cent) was recorded in plants treated with 10 ppm NAA having maximum fruit weight (2.9 kg) and yield (88 t/ha) (Singh *et al.*, 1997).

Hydroxy Ethyl Hydrazine

Plants forced with BOH produce fruit comparable in weight and shape to fruit from plants forcedwith ethephon. Harvest date in most cases was also similar to that for plants forced with ethephon and, at some times of the year, 6-10 days earlier than plants forced with NAA (Leeper, 1965; Yamane and Ito, 1969). Usually 3.0 kg of BOH in 1135 lit of water (Yamane and Ito, 1969) or 2000 1 ha^{-1} of a 2500 p.p.m. active ingredient (a.i.) solution was applied (Py *et al.*, 1987). For best results, BOH was applied late in the evening or at night. Because BOH degrades to pro-ice ethylene (Palmer *et al.*, 1967), it is assumed that this ethylene actually forces flower development. Higher concentrations BOH are phytotoxic, with symptoms developing soon after application (Anon., 1975}. BOH is not registered for use in the SA because the chemical was identified as potential carcinogen.

Ethyphon

Ethephon is less effective in forci·ng pineapple than are ethylene and acetylene but is ideal in environments where plant sensitivity to forcing is relatively high. An important advantage of ethephon under conditions ideal for forcing is that it can be applied with a boom sprayer during the day ir night with equal effectiveness. Ethephon locally applied in a 2-5 per cent (weight :vol-urea solution. Nitrogen sources such as ammonium nitrate and ammonium sulphate do not provide the same enhancement of ethephon activity as is provided by urea (Yamane and Ito, 1970). The mechanism by which urea enhances the activity of ethyphon is not known. Studies with radio labelled ethyphon show that urea does not increase absorption (Turnbull *et al.*, 1993, 1999). High solution of pH hastens the rate of ethyphon breakdown to ethylene, but urea only slightly increases solution pH. The addition of 2.91 of ethephon (39.5 per cent a.i.) to 17941 of 2.7 per cent urea-water solution raised the pH from 2.90 to 3.20 (Yamane and Ito, 1970). Glennie (1979) reported that a solution containing 480 mg l⁻¹ of ethephon had a pH of 2.7, while adding 5 per cent urea increased the solution pH to 3.0 and, at that pH, the solution half-life was 75,000 h.

Forcing with ethephon is more difficult when day temperatures are > 28°C (Glennie, 1979) or > 32°C (Wassman, 1991) and in tropical regions with high average temperatures. There are two approaches to improving the effectiveness of ethephon. In the first, the amounts of ethephon applied can be increased. Under idea! conditions, 0.5 kg ethephon a.i. is applied in 2000-3000 1 ha⁻¹ of a urea-water solution (Bartholomew and Criley, 1983; Py et al., 1987). When leaf nitrogen is high, up to 6 kg ha⁻¹ may be required (Guyot and Py, 1970). Higher amounts have been reported to reduce fruit weight (Dalldorf, 1979; Iglesias, 1979); however, others (Yamane and Ito, 1969; Guyot and Py, 1970) found no reduction in fruit weight or length, or other evidence of phytotoxicity when more than 16 kg ha⁻¹ was applied. However, Yamane and Ito (1969) found that slice recovery declined significantly at 13.4 and 17.9 kg ethephon ha/ha and sucker numbers declined significantly when 9 kg ha/ha or more was applied.

The second way of improving the effectiveness of ethephon is to raise its pH, which, as discussed above, increases its rate of breakdown to ethylene. Cox (1979) showed that the half-life of ethephon at 25°C is 761 h at pH 5, but drops to 12 h at pH 9. Glennie (1979) showed that, at this temperature at pH 3, the half-life was 75,000 h. At pH 9.2 at the temperatures experienced in Australian fields, ethylene release is rapid, and this has permitted relatively low rates of ethephon use. However, when temperatures are high, this high breakdown rate may be a disadvantage and sometimes, when field conditions are adverse, better inductions can be obtained with a solution pH of only 7. (Sinclair, 1991). Application between dusk and dawn, when temperatures are cooler and stomata are open, increases the effectiveness of ethephon, presumably because ethylene released by ethephon decomposition outside the plant is absorbed through the stomata. Two applications spaced 4-8 days apart may be required to force a high percentage of plants.

The almost universal induction of flowering with a growth regulator is almost unique to pineapple. Where the climate is favourable, the commercial exploitation of this response makes it possible to produce fruit in whole year. Year-round forcing and harvesting of pineapple make efficient use of available labour and provide a steady supply of pineapple to the cannery and the fresh-fruit market. Forced induction reduces variability in time of fruit maturation within a field, such that most of the fruits can be harvested within a few weeks. This period can be shortened by application of ethephon to mature green fruits to enhance fruit shell colour. While pineapple is generally easy to force, plant sensitivity to forcing varies considerably. It is important to understand how plant sensitivity varies, because such knowledge can help to increase forcing success.

Floral induction in pineapple by selective chemicals to obtain scheduled fruiting is now widely practised. Ethylene was found to be the active ingredient in the smoke of the greenhouse at Azore islands, that induced flowering (Rodriguez, 1932). Other chemicals such as acetylene, generated by calcium carbide (Kerns, 1935), ethylene applied as a gas (Cooper and Reese, 1941) and sodium salt of naphthalene acetic acid (NAA) showed usefulness to induce flowering in pineapple (Clark and Kerns, 1942). Betahydroxy-ethylhydrazine (BOH) in water sprays was also found effective in inducing flowering under certain conditions (Gowing and Leeper, 1955). Ethephon (2-chloro-ehylphosphonic acid) has also been found effective (Cooke and Randall, 1968; Py and Guyot, 1970; Wee and Ng, 1971). These have been widely used since the past 15-30 years with suitable modifications in respect of dosage, time and stage of application, depending upon the climate and local needs.

In areas where warm summers and frequent rains produce rapid plant growth, NAA and BOH have not been always effective on Smooth Cayenne (Py and Silvy, 1964; Py, 1963; Py and Barbier, 1966) or on Singapore Spanish (Wee and Ng, 1970). On the other hand, ethylene applied as a water spray has been successfully used in warm and wet tropics, where suitable equipment has been developed.

Acetylene-generating calcium carbide is also used extensively in many warm, rainy areas. Yet, it has frequently failed to induce flowering in a large proportion of plants in Smooth Cayenne (Py and Silvy, 1954; Py and Barbier, 1966; Py and Guyot, 1970), but occasionally it was partially effective on Singapore Spanish and Sarawak (Wee and Ng, 1968; 1970; 1971). Repeated day-time spraying of acetylene solution on Smooth Cayenne was found more effective than single application (Py and Silvy, 1954; Py and Barbier, 1966).

Meagre information is available on the time of application. Py and Barbier (1966) found a single application of BOH or calcium carbide either dry or wet at the centre of the Smooth Cayenne plants, and found it more effective at 4.00 a.m. than at 10.00 a.m. Aldrich and

Nakasone (1975) reported 100 per cent flowering by the application of calcium carbide (dry) as powder or with water at the core of the plants. The effective concentration varied between 0.5 per cent and 2.0 per cent and the best time of application was from 8.00 p.m. to 9.00 a.m.

At the Golden Circle Cannery in Australia several studies were made. Application of 2 rounds of sodium salt of cc-naphthaleneacetic acid (Sodium NAA) at 10 ppm at the rate of 2 fluid ounces per plant or sucker in the heart of the plant effected good flowering. Second dose of sodium NAA was given 14-21 days after the first. Another study used granular carbide for spot gassing, followed by 10 ppm sodium NAA within a week (carbide plus sodium NAA). In the third, 2 rounds of ethylene were applied with specially equipped boom-sprayers. Some basic differences exist between the 3 types of forcing agents. Carbide plus sodium NAA treated plants tend to throw well-shaped fruits with some slips. But this method is suitable only for small plantings or where only a limited number of plants are to be treated individually. Two rounds of sodium NAA method is the simplest but treated plants tend to throw slightly smaller fruits and fruits may be more tapered than normal. In addition, slip production is reduced (either a desirable or undesirable characteristic, depending upon whether slips are needed or not). Two rounds of ethylene method is suitable only to specially designed boom sprayers. Ethylene, for best results, must be applied in late afternoon or at dusk. Its advantages are well-shaped fruits, weighing slightly more than NAA forced fruits, and higher slip and sucker production. However, preliminary cannery studies showed no marked differences in recovery, which would favour one method over the another. Finally, the choice of the method is up to the individual growers.

In India, intensive efforts have been made since 1970 to select right chemicals, suitable concentration and time of application for flower induction in Kew pineapple. Application of NAA and NAA-based compounds like Planofix and Celemone has been reported to be very effective in inducing flowering (Das, 1964; Balakrishnan *et al.*, 1978a; Maity and Sambui, 1980). According to Burg and Burg (1966), flowering is a consequence of the stimulation of the ethylene biogenesis by about a week following auxin treatment. But disadvantages with NAA are delayed fruit maturity (Das, 1964) and inconsistent flower induction from season to season. Calcium carbide had resulted in off-season flowering in June to the extent of 93 per cent in Chethalli (Coorg). Small pellets of carbide at the rate of 1 g/plant were placed at the heart of the plant; acetylene was released when the product came in contact with water (Singh and Rameshwar, 1976). In Assam, 76 per cent flowering was achieved with calcium carbide (Das *et al.*, 1965). Ethephon (usually under the trade name of Ethrel) was found to induce more uniform flowering as compared to NAA or NAD (naphthalene acetamide) (Randhawa *et al.*, 1970). The ethephon concentration at 100 ppm was most effective. As treatment with 50 ml of 100 ppm Ethrel seemed to be costly, efforts were made at the Indian Institute of Horticultural Research, Bangalore, at Chethalli (Coorg) and at Thrissur, to lower concentrations of Ethrel. It was possible to induce more than 90 per cent flowering with 25 ppm of ethephon in combination with 2 per cent urea and 0.04 per cent sodium carbonate (Dass *et al.*, 1975; Balakrishnan *et al.*, 1978 a).

Urea helps in better absorption of any organic compound into the plant system and sodium carbonate increases release of ethylene by increasing pH of the solution.

Investigations carried out at the Pineapple Research Station at Mohitnagar in Jalpaiguri district of West Bengal have revealed that application of 50 ml solution per plant containing calcium carbide (20 g/ litre) or Ethrel (0.25 ml/ litre) causes flower induction.

Efficacy of growth regulators is influenced by the season. Trials were undertaken to find out suitable growth regulators for a particular season. Ethrel at 25 ppm along with 2 per cent urea + 0.04 per cent sodium carbonate was effective during all seasons. However, NAA was at a par during September to January. Still lower concentration of Ethrel, *i.e.* 10 ppm + 2.0 per cent urea + 0.04 per cent sodium carbonate, can induce 97 per cent flowering during dry months (March–May). At times or places where Ethrel is not available, NAA 10 ppm in combination with 2.0 per cent urea can induce more than 90 per cent flowering in Kew pineapple.

Based on the findings presented in Table 13, it would be advisable to use following growth regulators for different months for inducing flowering.

☆ *September-January*: NAA 10 ppm (Planofix 1 ml/ 4.5 litres of water)

☆ *March–May*: Ethrel 10 ppm (2.5 ml/100 litres of water) + 2 per cent urea + 0.04 per cent sodium carbonate

☆ *All months*: Ethrel 25 ppm (6.25 ml/100 litres of water) + 2 per cent urea + 0.04 per cent sodium carbonate

Table 13: Influence of Ethrel and NAA on Flower Induction in Different Seasons on Kew Pineapple

Sl.No.	Growth Regulators	Flower Induction (per cent) during			
		December–January	March–April	June–July	September–October
1.	Ethrel 25 ppm + 2 per cent urea + 0.04 per cent sodium carbonate	99.33	100.00	97.23	99.33
2.	Ethrel 10 ppm + 2 per cent urea + 0.04 per cent sodium carbonate	77.41	97.33	63.23	77.41
3.	NAA 1 0 ppm	98.23	50.66	88.66	98.23
4.	Control (water spray)	40.80	17.33	14.66	28.15

Stage of Growth for Chemical Application

Once the growth regulators or chemicals are selected, it is also very important to know the stage of the plant when it could be induced to flower. Though, growth regulators or chemicals selected can induce flowering at any stage of the plant growth, forcing plant to produce flowers at an early stage, but this reduces fruit size. When plants of optimum size are forced to flower at the right stage, better fruit size is obtained, without any adverse effects on the ratoon crop. Hence, stage of plant as denoted by the leaf number has been standardized. It was found that 35-39 active leaf stage (physiologically) was fit for treatment with 50 ml of 100 ppm Ethrel solution (Dass *et al.*, 1977). No beneficial effect was obtained with further increase in leaf number (Table 14). Optimum stage with regard to 'D' leaf was around 40-50 g.

Table 14: Effect of Ethephon at Different Leaf Number Levels on Yield Attributes of Kew Pineapple

No. of Leaves	Average Fruit Wt.	No. of Suckers/ Plant	No. of Slips/ Plant
25–29	1.077	0.53	0.58
30–34	1.413	0.77	1.47
35–39	1.789	1.24	3.41
40–44			
	1.780	1.09	3.57
45–49	1.914	1.11	3.64

For chemical induction of flowering in pineapple, the Golden Circle Cannery suggests that if average weight of 10 randomly collected 'D' leaves is 900 g, the plants are mature enough to produce good marketable fruits.

Method of Application

After preparing solution of required chemical or growth regulator, 50 ml of the solution should be poured into the heart of the plant. Efficacy of flower-inducing compound is reduced during rainy reason. Therefore, care should be taken that no rains are received for 24-36 hours after the application of the chemical. Otherwise application has to be repeated. Plants start flowering in 45-50 days after application.

Year-round Availability of Fruits

There is a possibility of spreading fruit harvest almost throughout the year by (i) taking up pineapple planting at regular intervals all round the year, (ii) using suckers and slips of different sizes and crowns as planting material and (iii) by the application of flower-inducing chemicals.

Studies carried out at the Pineapple Research Station of the Bidhan Chandra Krishi Vishwa Vidyalaya at Mohitnagar have revealed that in order to harvest fruits throughout the year, slips or suckers of pineapple should be planted from July to December, followed by application of Ethrel (0.25 ml/1) or calcium carbide (20 g/1), between 335 and 365 days after planting. The flower-inducing chemical should be applied to the heart of the plant between 7 p.m. and 8 p.m. in the evening and only when plants have at least 45-50 leaves. Application of chemicals at an interval of 7 days from April to November can ensure a steady harvest of fruits throughout the year (Aich, 1981).

Advantages of Flower Induction

1. Reduction in the cost of cultivation by restricting vegetative phase

2. Uniformity in harvest

3. Assured yield

4. Premium price to the grower through off-season crop

5. Regular supply of the fruits throughout the year to the canneries.

Improving Fruit Size and Quality

Application of NAA at a concentration of 300 ppm to the developing fruits 2 months after flowering resulted in maximum fruit size (Kwong-and Chiu, 1968). Poignant (1969) recorded an increase in fruit weight, yield and a reduction in total sugars by the application of sodium salt of NAA at 110, 155 and 200 ppm on the developing fruits at 12-16 weeks after emergence of inflorescence. The best time for spraying developing pineapple-fruits of Singapore Spanish with Planofix was 6 weeks after the emergence of inflorescence (Wee, 1971). It resulted in increased fruit weight, diameter and acidity and delayed fruit maturity. Experiments conducted at the Indian Institute of Horticultural Research, Bangalore, have shown that application of NAA at 200-300 ppm concentration 2-3 months after fruit set increases fruit size by 15-20 per cent (IIHR, Bangalore 1977). Similarly, 400 ppm of NAA applied 2 months before harvest also improved fruit weight (cited by Chadha, 1977). Roy (1981) suggested use of NAA (300 mg/litre) 45 days after flower emergence. The treated fruits showed an increase in weight and size and a reduction in total soluble solids and soluble sugar contents. Further, application of Ethrel (0.50 ml/1) on the NAA treated fruits at 120 days after flower emergence was found to improve fruit quality.

Ripening

Ethrel 2,000 ppm was found to induce uniform ripening which resulted in a single harvest of crop (cited by Chadha, 1977). Pre-harvest treatment of mature green pineapple variety Perola with Ethrel at 500-2,000 ppm resulted in development of uniform yellow colour in 8 days (Chunha *et al.*, 1980). Crochon *et al.* (1981) observed that application of Ethrel at 3 litres/ha, when fruits begin to colour, enabled complete harvesting 4 days later. While application of Ethrel 9 days before the harvest date at 5 litres/ha made fruits ready for harvest 5 days after treatment. Though, the appearance of the treated fruits was good, they were acidic and lacked flavour. On the other hand, 400 ppm of NAA solution applied to fruits one month before harvest, could delay maturity by 15 days (cited by Chadha, 1977). Das (1964) also experienced delayed fruit ripening with NAA treatments. These results may profitably be made use of depending upon the situation, either to postpone or to hasten harvest.

Late in the 19th century it was discovered that pineapple inflorescence development could be forced with smoke (Collins, 1960) and later research showed that the active ingredient in smoke was ethylene (Rodriquez, 1932). Work in Hawaii (Collins, 1960) showed that acetylene gas could also force inflorescence development of pineapple and that water-saturated solutions of acetylene or ethylene sprayed over plants could deliver the required quantity of either gas. Green leaf tissue is required for forcing with ethylene (Traub *et al.*, 1940), probably because the principal point of entry for gases is through the stomata. Calcium carbide, which releases acetylene on contact with water, also forces flowering. Calcium carbide must be placed in the centre of each plant, which precludes mechanization.

Later work showed that a variety of synthetic auxins also forced flower development of 'Smooth Cayenne' pineapple (Collins, 1960). The auxin used commercially was the sodium salt of naphthalene acetic acid (NAA) (Collins, 1960; Bartholomew and Criley, 1983). Almost 20 years after the commercial use of NAA began, Burg and Burg (1966) showed that auxins applied to pineapple stimulated natural ethylene production and speculated that the ethylene actually forced flower development.

Shortcomings with auxins and the technical and managerial difficulties associated with the use of acetylene and ethylene led to a continued search for better forcing agents. Over 700 compounds were screened during the 1950s but none were found to be better than NAA, acetylene or ethylene (Gowing and Leeper, 1959). At about the same time, Gowing and Leeper (1955) reported that p-hydroxyethylhydrazine (BOH) could force pineapple to flower. Palmer *et al.* (1967) showed that BOH decomposed to produce ethylene. It seems likely that the ethylene released from BOH induced flower development and the fact that BOH was most effective when applied at night is strong evidence for this.

Ethephon, which degrades to produce ethylene and may also stimulate ethylene production by the plant, was identified in the 1960s (Cooke and Randall, 1968). Extensive research (Bartholomew and Criley, 1983; Py *et al.*, 1987) has demonstrated the efficacy of ethephon as a forcing agent in a wide range of environments where pineapple is grown. While ethephon is less effective than ethylene in the warm tropics, its ease of use has caused it to be widely adopted for flowering of pineapple.

Plant Sensitivity to Forcing

Variation in plant sensitivity to forcing in different months of the year has been documented for 'Singapore Spanish' (Wee and Ng, 1968; Wee and Rao, 1977) and 'Smooth Cayenne' (Py *et al.*, 1987), and it is likely that other varieties also show such variation. As with natural induction, plants smaller than some minimum weight– usually somewhat less than about 1.0 kg fresh weight in subtropical regions–are not easily forced. However, forcing of plants of small weight is of little consequence to commercial growers because plants too small to be forced will not produce fruit of marketable size. Once a minimum plant size has been attained, plant sensitivity shows little variation with Increasing weight to some upper limit. Above this upper limit, which has not been well characterized, 'overly large' plants are reported to be more difficult to force than those considered to be of optimum size (Bartholomew and Criley, 1983; Py *et al.*, 1987). The cause of this reduced sensitivity is not known, but it may be related to a rapid growth rate, intense mutual shading or difficulty in delivering a growth regulator to susceptible tissues, or to more than one factor. As with very small plants, the issue is a minor one because growers would normally force induction before sensitivity begins to decline.

For plants of sufficient size, sensitivity to forcing reaches a maximum during the time of year when natural induction normally occurs. Outside this time, the sensitivity to forcing is determined primarily by the prevailing temperature, and this is true for both 'Singapore Spanish' and 'Smooth

Cayenne' pineapple (Wee and Rao, 1977; Bartholomew and Malezieux, 1994). In warm tropical regions, sensitivity to forcing, even during the time of natural induction, tends to be low because growing conditions are optimum and growth is rapid. For 'Smooth Cayenne', sensitivity usually decreases with increasing night temperature above about 25°C. In such environments, special care must be taken in order to ensure that a high percentage of plants in a field are forced. However, if vigorously growing plants can be forced, fruitlet numbers are high and, if conditions for fruit filling are favourable, large fruit are produced at harvest.

Outside the season when natural induction normally occurs, and particularly during the warmer months of summer, a reduced sensitivity to forcing that is short-term in nature is observed. This sensitivity is associated with the short-term effect of high day temperatures, usually above 32°C on a given day, although night temperatures may also be high. This type of reduced sensitivity, which is often seen during January and February in southern Queensland, determines not only whether plants can be forced, but also the 'quality' of the force (Wassman, 1991). 'Quality' of forcing is a subjective term used to characterize the intensity of the plant response to the plant-growth regulator as reflected in the time from forcing to the appearance of the inflorescence in the plant heart and the number of fruitlets initiated by plants of comparable size within a given environment. When the quality of the force is poor, fewer fruitlets per fruit will be initiated, fruit emergence is delayed and fruit weight and yield are markedly reduced Inflorescence and Fruit Development and Yield (Wassman, 1991). However, if the following day is cooler, sensitivity may be high and forcing is easily accomplished (Wassman,

1991). No data were found regarding what step in the induction process, *i.e.* fruitlet initiation or fruit development, is delayed on hot days when forcing is poor.

Variation Among Genotypes

As with natural induction, comparative studies show that clones of 'Smooth Cayenne' are less easily forced than are other varieties of pineapple (Wee and Ng, 1970; Py *et al.*, 1987). In Malaysia, 'Singapore Spanish' was the dominant canning variety for many years, mainly because it could be forced in this tropical region in all months of the year. 'Smooth Cayenne' pineapple grown on the peat soils of Malaysia is difficult to force in most months of the year. Breeding to improve plant and fruit characteristics has resulted in intervarietal hybrids of 'Smooth Cayenne' that are easier to force than 'Smooth Cayenne' and have higher yield and better flesh colour than 'Singapore Spanish' (Chan and Lee, 1985; Chan, 1986).

Environment

The environmental factors that promote natural induction also increase sensitivity to forcing. Variation in plant sensitivity has been detected when plants were forced with acetylene, NAA or ethephon, presumably because these plant-growth regulators have been widely used in a variety of environments. The relative effectiveness of acetylene, ethephon and NAA in forcing flowering varies and this variation is discussed below.

During the time of natural induction, there is little need for concern about the efficacy of the growth regulator. Induction of flowering outside this season can provide a severe test of forcing practices, especially for 'Smooth

Cayenne'. Sensitivity to forcing is enhanced by short days, cool temperatures and, where plants are likely to have reduced sensitivity, moderate nitrogen stress, established by withholding nitrogen fertilizer for 4-8 weeks prior to forcing. When forcing with NAA, the nitrogen content of the 'D' leaf should not exceed 1.6 per cent and a level of 1.3°C is recommended, especially if plants are expected to be difficult to force (Py *et al.*, 1957).

The effect of moderate water stress is apparently similar to that of moderate nitrogen stress (Evans, 1959), but no controlled studies were found. Severe water stress causes plants to be unresponsive to forcing, presumably because severe water stress probably causes all developmental processes to cease, thus rendering the plant insensitive to growth regulators.

When weather-shelter temperatures in subtropical regions exceed 28°C (Glennie, 1979) or 32°C (Wassman, 1991) during the day, forcing with ethephon is difficult. Under such conditions, leaf temperature may be 20°C higher than air temperature' (Bartholomew and Malezieux, 1994) and high leaf temperatures reduce leaf carbon-assimilation rates (Zhu *et al.*, 1999). In tropical countries, such as Cote d'Ivoire, Malaysia and Thailand, which have high average night temperatures, 'Smooth Cayenne' is difficult to force with ethephon and even more so with NAA (Py *et al.*, 1987). Average night temperatures of 25°C or greater can reduce the percentage of plants forced (Gonway, 1977) and the size of the fruit (Min and Bartholomew, 1997). Night temperatures of 25 and 30°C reduce natural ethylene production, CO_2, fixation at night into malic acid (Min and Bartholomew, 1997), total carbon assimilation (Zhu, 1996) and fruitlet number (Min and Bartholomew, 1997) of plants forced with ethephon. It seems likely that any environmental factor that reduces net assimilation, either by reducing photosynthesis or increasing respiration, at the time of forcing will reduce fruitlet number and thus fruit weight at harvest.

Other factors affect the number of florets initiated on a plant of a given weight, but for the most part these appear to be unrelated to variations in plant sensitivity to forcing. These effects are discussed under Inflorescence and Fruit Development and Yield below.

Prevention of Natural Flowering

When sensitivity to natural induction is high in subtropical regions, inhibiting natural induction can be a greater concern than is forced flowering. As the pineapple fresh-fruit industry has expanded, it has become common to force pineapple crops in all months of the year. As a result, natural induction before the scheduled forcing date has become a significant problem. In some areas, particularly subtropical regions, and in some years, precocious flowering may cause serious yield losses because it results in fruit that are too small or too few in number to be worth harvesting. Natural induction is less likely to be a problem in warm, humid, tropical environments. As noted previously, some intervarietal hybrids are more susceptible than 'Smooth Cayenne' to natural induction. As hybrids replace 'Smooth Cayenne' in fresh-fruit production, natural induction is likely to become more, rather than less, important.

1. Cultural Controls

During the time of year when natural induction normally occurs, cultural practices provide some degree of control over this process. These controls consist of ensuring that only relatively small, rapidly growing plants are in the field when natural induction is most likely to occur. Plants that are to be forced into flowering a few months after natural induction normally occurs should be grown with optimum levels of nutrients and water to minimize stresses and promote rapid vegetative growth.

2. Plant-growth Regulators

Synthetic auxins, such as NAA, force flower induction of pineapple at low concentrations while higher concentrations of these auxins delay it (Collins, 1960; Gowing and Lecpcr, 1960). Recent research with 2-(3-cholorophenoxypropionic acid (3-CPA) (Scott, 1993; Rebolledo-Martinez *et at.*, 1997; Rabie *et al.*, 2000; Rebolledo *et al.*, 2000) and certain tri-azole growth inhibitors (Min and Bartholomew, 1996). At present, no plant-growth regulator has been found that is completely effective in inhibiting natural flowering and no chemical is known to be registered for this process.

3. Genetic Engineering

Genetic engineering of a clone with reduced production of ethylene may also provide a method for controlling natural flowering. If anti-sense or trans-switch technology could be used to turn off the gene that codes for the ACC synthase associated with natural flower induction, then transformed plants would presumably not produce sufficient ethylene to initiate flowering. The key to this approach is to target the specific gene associated with natural flower initiation without affecting other ethylene-related processes in the plant.

4. Inflorescence and Fruit Development and Yield

The critical factors that determine fruit yield include the plants per unit area, the number of fruits harvested from those plants and the average fruit weight. At a given plant population density, the potential number of fruits available for harvest is determined by the number of plants that produce an inflorescence, the number of those fruits that mature and the number of fruit actually harvested. The cost of harvesting often dictates the percentage of fruit actually harvested, and losses commonly average 13 per cent, with a range of 5-30 per cent (Wassman, 1982). Even when harvesting fruits from fields of a few hectares, it can be uneconomic to have a harvesting crew visit the field with sufficient frequency to harvest every fruit that matures. The lowest losses are associated with fields with the least spread in maturity and requiring only one picking, while the greatest losses are associated with fields where up to ten picks are required. Ethephon can be used to hasten 'ripening' of the fruit, which may sharpen the harvest peak and increase recovery of fruit from the field, but it can also reduce average fruit quality and increase the variation in maturity and quality at the time of harvest.

Within a given environment, fruit weight at harvest is determined in large part by plant weight at forcing (Py and Lossois, 1962; Gaillard, 1969; Tan and Wee, 1973; Malezieux, 1988; Malezieux and Sebillotte, 1990a). However, other critical factors that affect average fruit weight per plant include the success of inflorescence induction, the quality of this physiological process 'in terms' of the number of fruitlets initiated per unit of plant weight and the fruit-filling process(es). At maturity, average fruit weight is determined by the number of fruitlets formed during inflorescence induction and the average fruitlet weight.

5. Plant Population Density and Plant Size

Effect of Plant Population Density on Average Fruit Size and Yield

The effects of planting density, and thus of competition for light, on average fruit weight and yield have been demonstrated many times (Bartholomew and Paull, 1986; Py *et al.*, 1987) and have been reconfirmed by recent studies Scott, 1992; Zhang, 1992; Christensen, 1994). Virtually all studies show that average fruit weight decreases approximately linearly with increasing planting density (Bartholomew and Paull, 1986), while yield increased linearly or in a curvilinear manner if densities were high enough). Fruit diameter (Treto *et al.*, 1974; Zhang, 1992) and fruit length (Norman, 1978) decreased as the planting density increased. Although yield components were not reported in most studies, over a moderate range of densities the decrease in fruit weight seems to be due mainly to a decrease in average fruitlet weight rather than in fruitlet number (Sanford, 1962; Pinon, 1981;). However, at very high densities, fruit are so small that it is likely that both fruitlet number and weight decrease. In recent studies (Scott, 1992; Zhang, 1992; Christensen, 1994), total yield increased with planting density to populations as high as 128,000 plants ha^{-1}, but the numbers of smaller fruit, which have a lower commercial value, increased at the highest densities. All varieties of pineapple examined respond similarly to increasing plant population density, although the slopes of the lines fitted to the data were different for 'Queen', 'Spanish' and 'Smooth Cayenne'. The differences in slope are probably due in part to differences in the efficiency of plants in producing a fruit, but differences such as those for 'Smooth Cayenne' grown at different locations are probably due largely to differences in environment or cultural practices.

Effect of Plant Population Density on Crop Duration and Spread in Maturity

The period from forcing to harvest is prolonged as planting density increases (Py *et al.*, 1987; Scott, 1992; Zhang, 1992; D. Christensen, 1994, personal communication). In Hawaii (Zhang, 1992), when plants were forced in September, no delay in maturity was found at plant population densities below 75,000 plant ha. In southern Queensland, days from induction in January, 1990 to peak harvest increased from 253 at a density of 46,100 plants ha^{-1} to 265 at a density of 80,700 plants ha^{-1}. A second study in 1994 had similar results, with days from induction to harvest increasing from 256 at 62,500 plants ha^{-1} to 272 at a density of 85,200 plants ha^{-1}. Christensen (1995) found that there was about a 1-day delay in mature for every 1000 plant ha^{-1} increase in density, though presumably there is a lower threshold where this effect is not observed.

Christensen (1995) reported that peak harvest was not determined at a density of 93,700 plant ha^{-1} because plants segregated into two populations, one of large, unshaded fruit that matured early and one of small, shaded fruit that matured much later. This segregation was probably due to the increased plant-to-plant variability that occurs at high plant population densities. Small variations in plant weight at planting become magnified as density increases because large plants overtop smaller ones. Development of small fruit buried in the canopy are delayed because such fruits have less sun exposure than do large fruit borne on large plants. Shading could retard development by lowering the average fruit temperature (Malezieux *et al.*, 1994) or by reducing the supply of assimilates allocated to the developing fruit. It is clear that fruit development is delayed, particularly where higher plant populations densities are used.

6. Plant-Weight–Fruit-Weight Relationships

Plant or leaf weight at the time of forcing and fruit weight at harvest are generally highly correlated for a given variety of pineapple within a given environment (Bartholomew and Paull, 1986; Py *et al.*, 1987; Malezieux, 1988, 1993; Zhang and Bartholomew, 1997). However, the relationship between plant weight at induction and fruit weight at harvest is complex and not always predictable. The strength of the relationship between plant or 'D'-leaf weight and fruit weight depends on the growing conditions, including climatic conditions, prevailing during a specific crop and hence it is not extrapolatable. Linford (1933) reported that the number of floret buds on plants from two fields was well correlated with stern and peduncle diameter, but less well correlated with stem weight and not significantly correlated with stem length. Further, the number of florets was greater for each stem-diameter class for plants from one field than from the other. Stem weight per floret was 1-65 g for plants from one field and 10.45 g from the second field, causing Linford (1933) to suggest that the factors that determine floret numbers may not be proportional to the plant's ability to earn its fruitlets to maturity.

In regions near the equator, where the environment is relatively uniform throughout the year, the correlation between plant or 'D'-leaf weight at forcing and fruit weight at harvest might be expected to be high during most months of the year. However, in Cote d'Ivoire, within a particular field variability in fruit weight might or might not be well correlated with plant weight at forcing (Malezieux, 1988). In these equatorial regions, it seems likely that the primary effect of plant weight is to determine fruitlet number rather than fruitlet weight at harvest (Malezieux, 1988). In an experiment where plants of different ages, and consequently plant weights, were forced at the same date (Malezieux and Sebillotte, I990a), the increase in plant weight at forcing was associated with an increase in fruitlets per fruit. The leaf-area index (LAI) in this experiment ranged from 2.0 to 10. The number of fruitlets in this experiment was linearly correlated with plant growth during the month following forcing.

A positive and significant relationship between plant weight and fruitlet number was also shown in an experiment in Cote d'Ivoire, where plots planted monthly where forced systematically at 8 months (Malezieux and Sebillotte, 1990b). Part of the residual variation might be related to climatic conditions in the month following forcing, because drought and low radiation reduced the expected number of fruitlets. The fact that there was no direct relationship between plant weight at forcing and fruitlet weight at harvest in these experiments might be related to the fact that fruitlet filling is the result of the balance between the source (whole-plant capacity to provide assimilates for the fruit) and the sink (number for fruitlets to be filled). Statistical relationships between plant weight at forcing and fruit weight at harvest may also be established. In Cote d'lvoire, a linear relationship was found for the data obtained from the monthly-planting trial previously referred to (Malezieux, 1993;). Part of the residual variation may be explained by climatic conditions after forcing, which influence fruitlet number and fruitlet filling. Data points that fall below the regression line are due to inadequate fruit filling for plantings made in May, June and July, because of drought and low irradiance during fruit development, while data points located above the line are due to good growing conditions during fruit development for plantings made in January and February. This relationship is not universal, but depends on a variety of factors including the climatic conditions prevailing after forcing the mineral status of the plant and the quality of pest and disease control.

In a time-of-planting trial conducted in Queensland, Australia (Sinclair, 1992b), which included multiple dates of forcing, the relationship between plant weight at forcing and fruit weight at harvest was not significant if no account was taken of season of induction. Seasonal influence was a major determinant of fruit weight in this experiment. Fruit weight was mainly determined by climatic factors–primarily temperature -that occurred during flower induction and fruit development rather than plant weight at forcing (Sinclair, 1992b). In this trial, plants weighing 3.0 to almost 4.5 kg that were forced in autumn had low (1.0 kg) fruit weight at harvest because fruit development occurred during the winter. Plants weighing only about 2.5 kg that were forced in spring produced fruit that had a fresh weight of about 2.0 kg at harvest.

7. Genetic Effects

Some genetic variation in the length of the fruit development period exists within hybrids (Chan and Lee, 2000), though it was apparently not great enough to significantly shorten time from forcing to fruit maturity. Variation in length of the fruit development period has also been found in 'Smooth Cayenne'. In Queensland, a comparison of four clones indicated that 'Smooth Cayenne' done 'CIO' matured approximately 7 days earlier than the 'Smooth Cayenne' clone 'F180'. The Pineapple Research Institute of Hawaii hybrid 53-116 matured approximately 14 days later than 'CIO'.

8. Fruit Enlargement

Synthetic auxins such as NAA, β-naphthoxyacetic acidicotic acid and 3-CPA applied after inflorescence emergence, and most often after flowering increase fruit weight at harvest (Bartholomew and Criley, 1983), and fruit yields can be increased by up to 40 per cent (Williams, 1987). The available literature indicates that only 3-CPA was used on a commercial scale. A possible added advantage of 3-CPA is that it reduces crown size, which is a particular benefit for fresh-fruit production. The hazards of growth regulator-induced fruit enlargement is delayed harvest, decreased shelf-life, crown damage, decreased total soluble solids (TSS) pale flesh colour, increased incidence of disease and increased numbers of green-shell ripe fruit (Bartholomew and Criley1983; Williams, 1987). The risks associaed with the use of 3-CPA appear to out weigh its benefits and, as a consequence, it is not registered for use in Hawaii and has been banned from use by growers providing fruit for the cannery in Queensland. However, it is still used to enlarge fruit of 'Queen' pineapple in Malaysia.

Chapter 9
Pests, Diseases, Plant and Fruit Abnormalities and their Management

Introduction

The pineapple plant is most productive under a xerophytic environment where low moisture is supplemented by irrigation in well-drained soils. The adventitious roots arising from the lower portion of the pineapple stem are the only ones that become soil roots. Once the root system is damaged or destroyed, it does not regenerate significantly.

Indices for pineapple-plant diseases are the proportion of the plant population affected (incidence) and the effect of disease on each plant (severity). Severity may range from a reduction in growth rate, as indicated by a reduced plant size or weight, to a reduced fruit yield. In order to understand the epidemiology of pineapple-plant diseases and to develop management strategies, the probability of disease occurrence (frequency) and the level of disease occurring (incidence and severity) must be considered. Factors of importance are the presence or absence of the causal organism, susceptibility of cultivars and optimum environmental conditions.

Pests and their Management

Nematodes

Four species of nematodes have been associated most frequently with, and caused the most damage to, pineapple: the root-knot nematodes, *Meloidogyne javanica* (Treub) Chitwood) and *Meloidogyne incognita* (Kofoid and White) Chitwood), the reniform nematode, *Rotylenchulus reniform* (Linford and Oliveia) and the root-lesion nematode, *Pratylenchus brachyurus* (Godfrey Filipjev and Schuurmans stekhoven) (Caswell *et al.*, 1990).

Root-knot Nematodes

The most obvious symptom of root-knot nematodes, *M. javanica* and *M. incognita,* on pineapple is the terminal club-shaped gall resulting from infection of the root tip. Less obvious symptoms include stunting of plants and water stress, with the terminally galled root resulting in poor plant anchorage. Nematode egg masses survive for relatively short periods (hours) in desiccated soils. Egg masses in galls may survive several days. Juveniles may survive several weeks to years in desiccated soils. Second-stage juveniles infect the pineapple root tip and become sedentary after 2-3 days. Vermiform males and saccate, sedentary females go through several moults. Surviving nematodes can tolerate a wide range of soil temperatures and pH.

Reniform Nematode

The reniform nematode, like the root-knot nematode, causes stunting of plant growth, with infected plants appearing to be under water stress, much the same as in drought, mealybug wilt or root rot. Symptoms are most severe in ratoon crops and may result in the total collapse and death of the plants. As with root-knot, above-ground symptoms are not diagnostic. In contrast to root-knot, however, pineapple plants infected with the reni-form nematode has excellent anchorage because of the lack of terminal galling. Infected primary roots continue to grow but secondary root growth is severely limited. Infected roots appear to have nodules, which are actually soil clinging to the gelatinous matrix of females embedded in the roots (Caswell and Apt, 1989). Reniform nematode eggs hatch when stimulated by root exudates of host plants. Second-stage juveniles in the soil undergo 3 moults without feeding, ending as either adult males or preadult females. The pre-adult females infect the root, where they establish sedentary feeding, become swollen mature adults and start producing eggs. The male does not feed.

Root-lesion Nematode

The infection sites of the root-lesion nematode, *P. brachyiirus*, are characterized by a black lesion that progresses along the root as the nematodes move for feeding. Secondary roots and root hairs are also destroyed. Initial inoculum comes from infested root fragments in the soil or infected roots on infested seed material. Once the plant is infected, the entire life cycle can be completed within the pineapple root. Reproduction is by mitotic parthenogenesis, with males being rare. Optimum soil temperatures are 25-30°C and populations do best in acid soils. In the highly acid Ivory Coast soils, the root-lesion nematode displaces the root-knot nematode. A combination of root-lesion nematode and *Pythium* species results in greater damage than either alone (Guerout, 1975).

Bud Moth, Pineapple Borer

The bud moth (also known as pineapple borer, pineapple caterpillar), *Thecla basilides* (Geyer), is found throughout Central and South America wherever pineapples are grown (Py *er al.*, 1987; Sanches, 1999).

The adult stage of the bud moth deposits its eggs on the inflorescence prior to anthesis. The larva infests the fleshy parts of the bracts and feeds inside the developing inflorescence. Buds and open flowers are entered directly, with larva penetrating developing fruit and digging out holes of varying depths. This results in malformed fruit, which is unmarketable (Py *et al.*, 1987). In response to the action of the larvae, the pineapple fruit forms an amber-coloured gum (gummosis), which exudes and hardens on contact with the air. This resembles the resin exuded by pine trees (sometimes called 'resinous' by Brazilians). When secondary erections are due to *F. moniliforme* var. *subglutinans*, the exudate is more fluid, characteristic of fusariosis disease. The pathogens can penetrate the inflorescence without the bud-moth larva, but the bud moth makes it easier to do so. Adult bud moths may help disperse the pathogen when visiting healthy plants after visiting diseased ones. Following a 13-16-day feeding period, the larvae emerge and pupate in the leaf axils. Control with insecticides is relatively easy if flowering is uniformly induced with forcing agents. Predators exist in Trinidad, including the vespid wasp *Polistes rubiginosus* Lepeletier and a predator of larvae, *Heptamicra* sp. Another predator reported is *Metadontia curvidentata*. No organized biological-control campaign has been undertaken for the bud moth (Py *et al.*, 1987).

Fruit-piercing Moth

There are numerous species (> 90) of moths in the lepidopteran family Noctuidae in which the adult stage (*i.e.* moth) pierces many types of fruit with a specially adapted proboscis (Banziger, 1982). The well-known pest called the 'fruit piercing moth', which attacks pineapple, is *Eudocima* (- *Othreis*) *fullonia* (Clerck) It does not limit its adult feeding to pineapple and may attack numerous fruits (*e.g.* oranges, guavas, star fruit, mangoes, bananas, coffee, passion-fruit, litchi, etc.) if available (Waterhouse and Norris, 1987). It may be found in many areas where pineapple is grown (Australia, Asia, Hawaii). The larval stages of the moth do not feed on pineapple, and outside the Pacific Basin they attack at least 30 species of creepers belonging to the plant family *Menispermaceac*. Within the Pacific Basin, the larvae are normally found feeding on coral trees in the genus Erythrina (*Fabaceae*) (Cochereau, 1977). They may be found on a creeper, *Stephania foresteri,* which is now rare in New Caledonia. The larval stages feed on the foliage of their host plants, where the yellowish green eggs are also laid. An individual female moth may lay as many as 750 eggs, either singly or in batches of up to 100-200 eggs (Waterhouse and Norris, 1987). Several generations are possible during a year. Injury to fruit is caused by strongly sclerotized appendages (maxillae) at the end of the approximately 25.4 mm proboscis. The tips of the maxillae are equipped with a series of teeth and spines, which enable the adult moth to rasp a hole through the tough skin of the pineapple fruit. The act of piercing the fruit requires little time, from several seconds to a few minutes (Cochereau, 1977). Controlling *E. fiillonia* is difficult with pesticides normally being ineffective, since the larval stages of the pest are not within the pineapple crop. Pesticides applied to the ripe fruit can also cause human-health concerns, due to the presence of toxic residues (Waterhouse and Norris, 1987). Several approaches have been tried for the control of fruit-piercing moth, but none are especially effective or, if so, they are impractical or expensive to implement. These include use of poison baits for control of the adult stage; confusing searching adults by masking fruit volatiles in orchards with smoke; use of potential repellents; orchard sanitation to reduce quantities of decaying or fallen fruit; early harvest of fruit; hand-capturing and killing of moths; bagging of fruit; using floodlights to disrupt moth flight behaviour; and eradication of host plants (Baptist, 1944; Cochereau, 1972a; Banziger, 1982). Where natural enemies exist and conditions are favourable, biological control has proved effective in New Caledonia (Cochereau, 1972b, 1977). Natural enemies include para-sitoids and predators of eggs and larvae, as well as a fungal disease of the eggs (*Fusarium* sp.). Egg parasitoids in the genera *Ooencyrtus* (Hymenoptera: Encyrtidae) and *Trichogramma* (Hymenoptera: Tricho-grammatidae), as well as the larval-pupal parasitoid *Winthemia caledoniae* Mesnil (Diptear: Tachinidae), are excellent candidates for introductions into areas where biological control of the pest is poor. Additionally, reduction in the abundance of the larval host plants could help suppress adult numbers within an area (Waterhouse and Norris, 1987).

Giant Moth Borer

The giant moth borer, *Castniomera* (= *Castnia*) *licus* (Drury) (Lepidoptcra: Castniidae), is found in South America (*e.g.* Brazil), where it is a minor pest of little economic consequence on pineapple (Collins, I960). The mature larvae of this species live up to their common name and are about 7.6 cm in length. Larvae may enter the stern of the pineapple plant and burrow vertically into the peduncle, which supports the fruit. This action disrupts fruit production (Collins, 1960). Adult moths have a wing spread of about 12.7 cm.

Mites

The Pineapple Red Mite

The pineapple red mite (also known as red spider or false spider mite), *Dolicho-tetranychus* (= *Stigmacus*) *floridanus* (Banks) (Acarina; Tenuipalpidae), is the largest mite found on pineapple and is conspicuous *en masse* because of its bright orange to red. According to Jeppson *et al.* (1975), it only occurs on pineapple and is found in Florida, Cuba, Puerto Rico, Panama, Honduras, Mexico, Central America, Hawaii, the Philippine Islands, Japan, Okinawa and Java. The adult mite is approximately 0.3-0.4 mm long and O.lmm wide. When present on the plant, the mite is always found on the white basal portion of the leaves, where it feeds, particularly on the crown. When pineapple red-mite populations build up under dry conditions, the rnites are most commonly on the basal leaves of the crown and on stored seed material (Petty, 1975,1978c).

The Blister Mite

The blister mite (also called pineapple fruit mite), Phyllocoptruta (= Vasates) saki-murae Kiefer (Acarina: Eriophyidae), is reportedly the smallest mite (0.1 mm long and 0.033 mm wide) found on pineapple in Hawaii (Carter, 1967). Individuals are chalky in colour and only have two pairs of legs located near the head. They may be found on detached crowns that are stored for planting. They originate from prior infestations on the ripe fruit from which the crowns were derived. They normally disappear after the crowns are planted, but may be found later on fruit after the flat-eye stage of fruit development (Carter, 1967). Jeppson *et al.* (1975) suggest that the mite originated in South America.

The Pineapple Mite

The pineapple mite, Schizotetranychus asparagi (Oudemans) (Acarina: Tetra-nychidae), is widely distributed and has been recorded in Hawaii, continental USA, Germany, Portugal, The Netherlands and Puerto Rico (Jeppson *et al.,* 1975). In colder climates, it may be found on asparagus ferns grown in greenhouses or lathhouses. In pineapple-production areas, it may frequently cause severe damage to recently established plants in the field. Plants that are infested in the early stages remain small and fruit production is either curtailed or non-existent. Heavily infested plants may die before producing fruit. The best management action is to plant only mite-free seed-plant material (Jeppson *et al.,* 1975).

The Pineapple Tarsonemid Mite

The pineapple tarsonemid mite (also known as pineapple mite, pineapple fruit mite, pineapple false spider mite),

Stencotarsonemus ananas (Tryon) (Acarina: Tarsonemidae), may be found infesting pineapple later in the plant's phenological cycle.

Mealybugs and Ants

The mealybug species associated with mealybug wilt are commonly found on pineapple seed material in most major production areas of the world. In order of importance worldwide, they are: pink pineapple mealybug, *Dysmicoccus brevipes* (Cockerell), grey pineapple mealybug, *D. neobrevipes* Beardsley, and long-tailed mealybug, *Pseudococcus adonidum* L. (Targioni-Tozzetti)). Insufficient mealybug control can lead to whole pineapple plantings being lost due to mealybug wilt, resulting in reduction of fruit production (Carter, 1933). Plantings may be infested with only one mealybug species or multiple species (Gonzalez-Hernandez *et al.*, 1999). These species are not equally distributed worldwide. The pink pineapple mealybug may be found in all major pineapple-growing areas. The pink pineapple mealybug is commonly found on the lateral roots of the pineapple plant just below soil level. It can also be found on the aerial parts of the plant, mainly in the leaf axils and on the developing fruit. In contrast, the grey pineapple mealybug is never found on the pineapple roots, but may overlap the distribution of the pink pineapple mealybug on the aerial portions of the plant. The male grey pineapple mealybugs spin white, silky cocoons before becoming small-winged insects, which seek females to mate. Mealybugs feed on plant sap in the phloem of their host plants. They produce honeydew (sweet, sticky liquid) as a by-product of their feeding. The honeydew often accumulates in large quantities around groups of mealybugs and may support the growth of sooty mould, *Capnodium* sp. Adult mealybugs are elliptical-shaped (top view), soft-skinned insects with waxy secretions, which give their body surfaces a chalky appearance. They also have white, waxy filaments of various lengths (depending on the species) extending from the lateral margins of their bodies. Although capable of movement, these insects are normally quiescent and congregate together in groups (*e.g.* > 20 individual). The first-stage (or first-instar) crawlers (0.6-0.7 mm) are typically the most active stage, and they move around the plant host seeking a place to settle down to feed. After settling down, they do not normally move any great distance, unless disturbed or relocated by ant species (*e.g.* big-headed ant) that tend them for their honeydew.

Mealybug wilt of pineapple, with its leaf-tip dieback and plant yellowing and reddening, is a symptom associated with the feeding of mealybugs. The actual cause has not been conclusively demonstrated, but one or two closteroviruses have been implicated (Sether and Hu, 1999). Mealybug wilt is a universal problem; the only exception may be in parts of Thailand, where wilt does not occur even though mealybugs are present. Mealybug wilt is clearly one of the most destructive diseases of pineapple plants, and field controls must be initiated during the fallow period and continued to harvest. High mealybug populations are required to cause wilt.

Ants are necessary for populations of mealybugs to develop and reproduce in pineapple fields, where mealybug parasitoids and predators are present. At least three species of ants are associated with mealybugs the big-headed ant, the Argentine ant and the fire ant (Rohrbach and Schmitt, 1994). Two other species–the long-legged crazy ant, *Anoplolepis longipes* Jardon), and the white-footed ant,

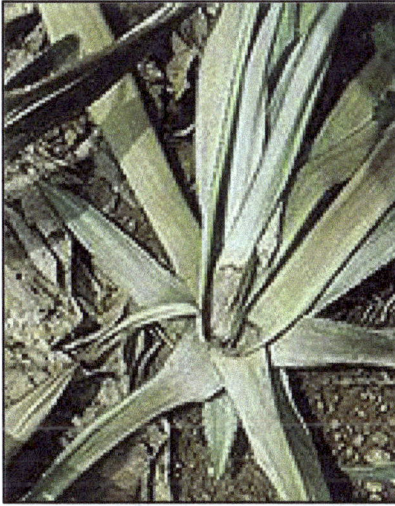

Figure 15: Pineapple Mealy Bug

Technomyrmex albipes (Fr. Smith)–are clearly associated with mealybugs in pineapple fields and, although not demonstrated to be associated with wilt, may have a role because they clearly tend mealybugs The ant association with mealybugs involves protection from predation and parasitism, removal of excess honeydew, which increases mealybug mortality and movement of the mealybugs into new areas. Preventing the establishment of new ant colonies in new plantings is critical to preventing mealybug wilt (Rohrbach and Schmitt, 1994).

White Grubs

The white grubs (*i.e.* larval stage) of several beetle species in the family Scarabaeidae commonly infest the roots of pineapple plants. Scarab species reported feeding on pineapple roots include, in Australia: the southern one-year canegrub (also known as rugulose canegrub, nambour

canegrub), *Antitrogus mussoni* (Blackburn), Christmas
beetle, *Anoplognathus porosus* (Dalman), Rhopaea canegrub,
Rhopaca magnicornis Blackburn, squamulata canegrub,
Lepidiota squamulata Waterhouse (= *Lepidiota darwini*
Blackburn, *Lepidiota leai* Blackburn, *Lepidiota rugosipennis*
Lea), noxia canegrub, *Lepidiota noxia* Britton, and *Lepidiota
gibbifrons* Britton (Waite, 1993); in South Africa: *Adoretus
ictericus* Burmeister, *Adoratus tessulatus* Burmeister,
Trochalus politus Moser and *Macrophylla ciliate* Herbst; and
in Hawaii: Chinese rose beetle, *Adoretus sinicus* Burmeister,
and *Anomala* beetle, *Anomala orientalis* Waterhouse (Carter,
1967), The species *Heteronychus arator* (Fabricius) is found
in Africa and Australia, where it is referred to by the
common names black maize beetle (Petty, 1976a) and
African black beetle (Waite, 1993), respectively. These
species above vary in the levels of damage they cause to
pineapple. Additional scarab species that attack pineapple
may also exist in these areas and other locations where
pineapple is grown. Scarabs are not limited to pineapple in
their feeding habits and may attack a wide range of plants.
The various species of beetles can be separated in both the
larval and adult stages *by* morphological characters on the
body (Carter, 1967).

Many species do not feed on pineapple plants as adults.
However, exceptions do exist, such as the adult stage of *H.
arator* which occurs in Australia, New Zealand and South
Africa (Waite, 1993; Petty, 1977a) and bores into the lower
sterns of the pineappJe plant. Adults of *A. sinicus* in Hawaii
may riddle or completely destroy pineapple leaves, while
the larvae rarely attack the roots (Carter, *1967*), Fortunately,
adult *A. sinicus* infestations are typically spotty in an area.
On the other hand, adults of *A. orientalis* typically remain

in the soil and lay their eggs in the vicinity of where they developed (Carter, 1967).

Adult scarab females are free-flying and choose the locations where they will lay their eggs in moist soil. Egg deposition preferences for soil conditions and type vary among scarab species (Waite, 1993). Eggs are oval in shape and, after hatching, the first-instar larvae feed on organic matter in the soil. Older scarab larvae develop within the soil among the roots of their host plants (*e.g.* pineapple). They feed upon organic matter within the soil as well. Although white grubs are not immobile, they do not disperse far from where the eggs were laid. White grubs are easily identified by their white or ivory-coloured, 'C'-shaped bodies, which are soft and plump. The posterior quarter to third of the larval abdomen is commonly a dark blue-grey colour, due to the contents of the digestive system. Grubs have three pairs of legs near their anterior end and a tan to dark brown head capsule (Waite, 1993). They may injure pineapple plants by; (i) feeding on the roots, which interferes with nutrient and water uptake and transport (Carter, 1967; Waite, 1993); (ii) weakening or destroying the roots that anchor the plants in the soil (Waite, 1993); and (iii) wounding plant tissues, which enables secondary plant pathogens to enter the plant (Carter, 1967). If infestations are severe, a crop may be lost, especially in the ratoon crop (Waite, 1993). The length of the scarab developmental cycle varies among species and climatic conditions, but generally they grow slowly compared with most insect pests and may require 1-2 years to complete development fo the adult stage (Waite, 1993). Recognition of white-grub infestations is difficult until significant injury to pineapple plants becomes obvious, commonly in the ratoon crop (Waite, 1993). Plants may become stunted, wilted and chlorotic. Severely affected plants are easy to pull out of the ground (Waite, 1993). Additionally, pathogens, such as *Pythium* fungus and root-knot nematode, may infect the plant (Carter, 1967). Areas designated for pineapple plantings should be inspected for the presence of white grubs prior to planting the seed crop. Larvae in the soil may be uncovered using a spade or found during cultivation of the soil (Waite, 1993). Adult beetles may be monitored using light traps. Thorough cultivation of the soil will reduce white-grub populations. A preplant soil treatment with long-term residual activity is appropriate for areas where white grubs are historically a recurring problem (Waite, 1993). Given the long production cycle of pineapple (*i.e.* seed crop and ratoon crops), the long-term effectiveness of chemical soil treatments is limited. Discoveries of white-grub infestations after planting are problematic because of the difficulty in controlling them. Delivery of chemicals to the insects is a challenge. Natural enemies of these pests do exist (insect predators, parasitoids and pathogens, as well as birds, toads, wild pigs and rodents), but the *levels* of control are not typically adequate (Carter, 1967).

Scales

Pineapple scale, *D. bromeliea* varies in its impact on pineapple. In some places (*e.g.* Australia), it does not typically reduce fruit yield directly, but affects fruit appearance so that the value is reduced (Waite, 1993). In other places (*e.g.* Hawaii, South Africa, etc.), high scale densities kill plants (Carter, 1967; Petty, 1978b; Py *et al.*, 1987). Scales normally occur on leaf undersides but may be found on the upper leaf surfaces if plants are shaded. Yellow spots may develop on leaves when scale densities are low. Scales have their greatest impact on the ratoon crops, where suckers and fruit may be damaged if shaded. Heavy pineapple scale infestations may weaken and stunt plants, producing a grey appearance and foliage dieback. Scale infestations may be found on the bottom eyes of mature fruit and all over lodged ratoon fruit (Waite, 1993). Cracks between fruitlets may develop when fruit are highly infested (Linford *et al.*, 1949; Py *et al.*, 1987). Volunteer plants that emerge next to ratoon plants may exhibit high scale densities.

In addition to chemical controls, this pest may be biologically controlled by natural enemies (Waite, 1993). Tiny wasps, including *Aphytis chrysomphali* (Mercet), *Aphytis diaspidis* (Howard) and *Aspidiotiphagus citrinus* (Craw) (Hymenoptera: Aphelinidae), parasitize the scales, resulting in scale death (Zimmerman, 1948). Ladybirds, such as *Rhyzobius lophanthae* Blasid. and *Telsimis nitida* Chapin (Coleoptera: Coccinellidae), also prey upon the scales (Carter, 1967; Waite, 1993).

Thrips

As many as 39 species of thrips have been reported worldwide in and around pineapple fields (Sakimura, 1937; Carter, 1939; Petty, 1978d). Wind currents may carry many individuals of these species from their host plants (*e.g.* domestic crops, ornamentals, weeds) into adjacent pineapple plantings (Carter, 1939). However, most of the thrips do not normally feed or reproduce on pineapple. Sakimura (1966) only reports six thrips species commonly living within Hawaiian pineapple plantings. These include the onion (or potato) thrips, *T. tabaci* (plant feeder), leaf thrips, *Frankliniella sulphurea* Schmutz (flower feeder), Thrips hawaiiensis (Morgan) (flower feeder), *Aleurodothrips fasciapennis* (Franklin) (predator on scales), *Haplothrips melaleucus* (Bagnall) (predator on scales) and *H gowdeyi* (Franklin) (flower feeder). Species reported as vectors of tomato spotted-wilt virus, which causes yellow spot on pineapple, are the onion thrips and the common blossom thrips (also known as kromnek thrips, cotton-bud thrips, yellow-blossom thrips), *Frankliniella schultzei* (Trybom) (Sakunura, 1963; Petty, 1978d). Pineapple yellow spot can be controlled by maintaining a weed-free production system to eliminate the weeds that serve as virus reservoirs. Adjusting the timing of operations of adjacent crops as well as pineapple to minimize the movement of thrips vectors into the pineapple fields aids in the reduction of inoculum (Rohrbach and Apt, 1986).

Souring Beetles

Small nitidulid beetles (c. 4.5-8.0 mm), known as souring beetles, sap beetles or dried-fruit beetles, are attracted to decomposing pineapple plant material (termed pineapple trash) following knock-down of the previous crop. The adult beetles are hard-bodied and dark brown (Hinton, 1945). Several different species may infest trash or overripe pineapple fruit, of which *Carpophilus humeralis* (F), *Carpophilus hemipterus* (L.) and *Haptoncus ocularis* (Fairm) are the most common (Carter, 1967; Py *et al.*, 1987). Fertile females may lay more than 1400 eggs and live as many as 115 days (Carter, 1967). Eggs usually hatch within 2 days after deposition. Hinton (1945) indicates that the life cycle of C. *humeralis* from egg to adult is about 21 days. While the larvae typically feed on decaying fruit, the adults may attack pineapple plants at every stage of growth (Hinton, 1945). They may congregate on seed plants placed in the field and feed on the exposed butts and starchy stalk material. However, the injury to the plant is not economically significant. Chang and Jensen (1974) have identified these beetles as being possible vectors of the fungus *Chalara paradoxa* (De Seynes) Sacc. (syn. *Thielaviopsis paradoxa* (De Seyn.) Hohn) (telemorph *Ceratocystis paradoxa* (Dade) C. Morcau) which causes black-rot disease. Souring beetles are more of a social nuisance than an agricultural one, because they often land on humans in the vicinity of knocked-down fields and fruiting pineapple plantings. This has been a problem in places such as Hawaii, where recreational and tourist activities (*e.g.* golf) are enjoyed near pineapple production areas.These beetles do not have a significant impact on pineapple production, it is not economically feasible or environmentally desirable to control them with pesticides. However, to reduce their nuisance factor, the parasitic wasp Cerchysiella (= *Zeteticontus utilis* Noyes (Hymenoptera: Encyrtidae) was collected in Israel and released in Hawaii in1977 to control the immature larval stages of the beetles that infest rotting pineapple trash and fruit (Funasaki *et al.,* 1988). The adult female parasitoid deposits her eggsinto beetle larvae and the parasitized larvae mummify (*i.e.* turn hard and stiff) after 9-11days. Fifteen days after egg deposition, an adult *C. utilis* emerges from the parasitized beetle.

Diseases and their Management

Butt Rot

Butt rot or 'top rot' of pineapple can be serious on pineapple 'seed materials' and occurs wherever pineapple is grown (Rohrbach, 1983). The causal fungus, *Ceratostomella paradoxa,* is widespread in the tropics on pineapple, coconut and other palms, sugar cane as 'pineapple disease', cacao as 'pod rot', and banana as 'black-head disease' on rhizomes, suckers and. roots, and as 'stem-end rot' on fruit (Dade. 1928). The symptoms of butt rot are a soft rot and blackening of the basal portion of the stem tissue of vegetative seed material. If infected seed material is kept wet, as in a pile of crowns, the infection may progress to rot the entire seed piece (stem and leaves) or even the entire pile. Severely rotted seed material is normally discarded prior to planting. Slightly to moderately infected seed material may be planted, but growth will be slow and plants will be stunted, due to loss of stem tissue, which contains carbohydrate reserves and the initial roots. When uncured or untreated seed material is planted in soils with high inoculum levels of *C. paradoxa,* butt rot levels may reach 100 per cent. Inoculum levels in pineapple soils in Hawaii varied from an average of 2630 propagules g^{-1} following field preparation to 280 propagules g^{-1} of soil at the end of the crop cycle. At planting, inoculum levels varied by field from a high of 12,969 to as low as 31 propagules g^{-1} of soil (Rashid, 1975). This diseases can be controlled by dipping plant material in 0.3 per cent Dithane Z-78 or by spraying on leaves. Copper fungicides should not be used in pineapple as they cause leaf scorching.

Fusarium Stem Rot

Fusarium stem rot is caused by the fungus *Fusarium subglutinans* (Wellenw. and Reinking) Nelson, Tousson and Marasas comb, nov,. O'Donnell *et al.* (1998) renamed the pathogen *Fusarium guttiforme* Nirenberg and O'Donnell based on DNA sequence analyses of members of the *Gibberelia fujikuroi* complex. Despite its obvious affinities with G. *fujikuroi*, no teleomorph has been reported for this pathogen. In Brazil, the disease causes major losses in the three major cultivars, Perola', 'Jupi' and 'Smooth Cayenne' (Rohrbach, 1983). Levels of plant infections van,' from 2 to 30 per cent (Laville, 1980). The disease is associated with the fruit-rot phase termed 'fusariosis'. Stern infections of seed materials occur at leaf bases, with resulting resetting and/or curvature of the plant, odue to portions of the stem being girdled or killed (Laville, 1980). Once the developing fruit is infected, secondary infections can occur on the developing slips or suckers. The infected seed material is then distributed to new planting areas, thus infesting new sites. Soils can remain infested for several months. Spread within infested fields is primarily by insects but may also be by wind (Laville, 1980). Free conidia of *Fusarium subglutinans* can survive for 6-13 weeks in soil, depending on *moisture* and temperature, with survival being highest in dry soils. Survival in infected pineapple tissue in soil is less than 10 months (Maffia, 1980). Optimum temperatures for growth are 25°C, with a range of 5~35°C (Camargo and Camargo, 1974).

Heart Rot or Stem Rot

The disease is caused by *Phytophthora cinnamomi* and *Phytophthora parasitica*. Infection due to P. *cinnamomi* is limited to areas of high rainfall and cool temperatures, although fungus is often present in soils of warmer and low rainfall areas. On the other hand, P. *parasitica* causes heart rot in warmer and somewhat drier areas. This organism is commonly seen in India. Poor physical condition of the soil and inadequate drainage are responsible for spread of the disease. It is frequently associated with alkaline soils but is not limited to them only. In this, green leaves turn yellowish-green and tips turn brown. The central whorl of leaves when affected will come out with a gentle pull Basal portion of the leaves shows signs of rotting and emits foul odour.

Leaf Spot

It occurs frequently in moist, warm, climate of eastern parts of India. Small water-soaked areas develop on leaves which gradually enlarge having pale-yellow colour and affected portions dry gradually. This disease is also caused by *Phytophthora* spp. (cited by Bose, 1985). Control measures are similar to heart rot Control measures include good drainage, proper selection of healthy planting material and prophylactic treatment of planting material with Dithane Z-78 (0.3 per cent) or Foltaf (0.4 per cent). Affected plantations can be sprayed with Dithane or Foltaf.

Root Rots

Root rots may be caused by *Phytophthora cinnamomi* and various *Pythiun* species, with *Pythium arrhenomanes* Drechs. as the most common (Klemmer and Nakano, 1964). Initial symptoms are a reduction or elimination of growth, with subsequent reddening of the leaves, the leaf margins turning yellow and eventually becoming necrotic. With *P. cinnamomi*, which causes heart and root rot, the root-rot phase results in reduced plant growth and yields and, in cooler environments, can result in a total loss of the ratoon crop. Root-rot symptom development is relatively slow in comparison with heart-rot symptoms. Disease from both pathogens is most severe when soils are cold and poorly

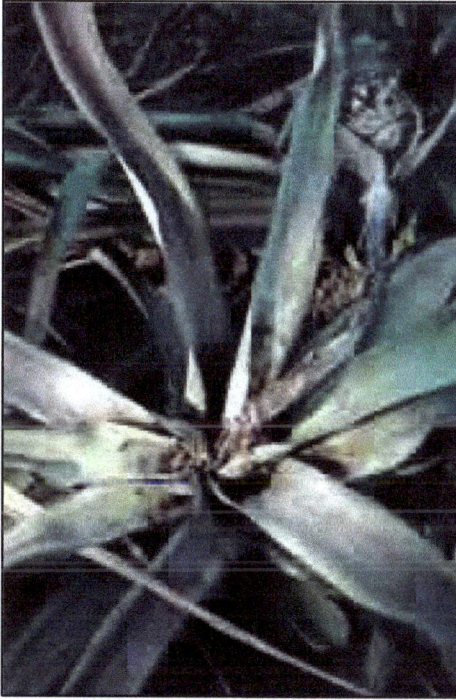

Figure 16: Heart Rot

drained. If environmental and soil conditions become dry following the infection period, affected plants may appear reddish, as if under severe drought stress. Plant anchorage in the soil is very poor following loss of roots.

Early Root-Health Management

With the discovery of the soil fumigant 1,3-dichloropropene, 1,2-dichloro-propane, (DD mixture) pineapple nematodes were easily and economically controlled during the early stages of pineapple plant growth (Carter, 1943; Keetch, 1979; Johnson and Feldmesser, 1987;

Figure 17: Symptoms of
***Phytophthora* Heart Rot**

Figure 18: Symptoms of
***Phytophthora* Root Rot**

Figure 19: Symptoms of *Pseudomonas* Heart Rot

Caswell and. Apt, 1989). Today, early nematode control is accomplished by clean fallow, preplant soil fumigation with dichloropropene at 224-336 1 ha^{-1} and postplant application of an approved nematocide (*e.g.* fenamiphos and oxamyl) by broadcast sprays or drip irrigation (Rohrbach and Apt, 1986; Caswell *et at*, 1990). Effective soil fumigation requires good plant-residue management and soil preparation. The discovery and use of the inexpensive DD control may have affected the development of other methods of nematode management for pineapple, such as cover crops, crop rotation and host-plant resistance. Crop rotations and resistance have been examined but not researched in depth or used Caswell and Apt, 1989; Caswell *et al.*, 1990). The root-knot, reniform and root-lesion nematodes have relatively large host ranges. Thus, crop rotations are of value only if crop susceptibilities are known.

Root rots are controlled by improving soil water management, including raised beds, deep cultivation and improving surface-water drainage. The fungicide fosetyl aluminium has shown good control of *P. cinnnamomi* root rot (Rohrbach and Schenck, 1985). The soil fumigant mixture of DD (Telone) was shown to reduce root rot caused by *P. arrhenamanes* (Anderson, 1966). Mealybug wilt is readily managed by controlling ants, which tend and protect mealybugs, with an approved insecticide bait (*e.g.* hydrarnethylnon) (Rohrbach *et al.*, 1988).

Fungal and Bacterial Heart Rots

Fungal Heart Rot

Fungal heart rots as well as root rot of pineapple are diseases associated with wet environmental conditions. *Penicellium cinnamomi* Rands requires cool conditions and heavy, wet, high-pH soils. Heart-rot mortality can range from 0 to100 per cent *c,* depending on the soil type, pH and rainfall. The economic impact of heart rot suits from a reduction in plant densities due to plant mortality. Heart rot can be caused by *Phytophthora nicotianae* B. de Haan van *parasitica* Dast. Waterh., frequently called *Phytophthora parasitica* Dast. in the pineapple literature, *P. cinnnamomi* Rands and *Phytophthora palmivora* (Butler) Butler. Heart-rot symptoms are the same, regardless of the *Phytophthora* species causing them. The most widely distributed species are *P. cinnamomi* and *P. nicotiana* B. de Haan var. *parasitica*

(Rohrbach, 1983; Rohrback and Apt, 1986). *P. palmivora* probably has a much more limited distribution (Boher, 1974; Rohrbach, 1983). Heart rot from *P. cinnamomi* is found under cooled conditions, such as the higher elevations and the cooler pineapple-production areas, where optimum soil temperatures for disease development are 19-25°C. An initial heart-rot symptom is the failure of the young leaves to elongate. Later symptoms are yellowing to bronzing of the young leaves, which may then lean to one side of the plant. A slight pull on the young symptomatic leaves will remove them from the plant, confirming the presence of the disease. *Phytophthora* infections are limited to the stem and basal white portion of the leaves. The primary inoculum of the three *Phytophthora* species is chlamydospores, either alone or in infested plant debris in the soil, where they can survive for years. Very little is known about the effects of soil moisture on infection. Infection by *P. nicotianae* var. *parasitica* is probably less dependent on high moisture than that by *P. cinnamomi*. High soil moisture (poor drainage) increases infection by *P. cinnamomi* but also reduces root growth. More recently captafol (Difolatan®), metalaxyl (Ridomil®) and fosetyl aluminium Allette®) have been used (Rohrbach and -Schenck, 1985). Currently, fosetyl aluminium is used very effectively as a preplant dip at rates of 2.24 kg active ingredient (a.i.) 935 $I^{-1.}$ Initial control from the preplant dip can be extended by foliar applications with rates of 6.72 kg ha^{-1} in 2805 1 of water at intervals of 3-6 months. Because fosetyl aluminium acts systemically in the pineapple plant, excellent control of *P. cinnamomi* root rot can be obtained (Rohrbach and Schenk, 1985; Rohrbach and Apt, 1986).

Crowns or slips from plants with symptoms of fruit collapse or from an area having high a incidence of fruit collapse should not be used as seed material. Mechanical leaf damage, such as occurs when entering a field for crop logging, should be minimized during periods of susceptibility and when low levels of disease are present. Partial control of bacterial heart rot has been obtained with miticides (*e.g.* endosulphan) and insecticides in the Philippines. Bordeaux mixture has resulted in variable control.

In subtropical climates, where the disease is a problem, tin- resistant 'Smooth Cayenne' cultivar might be used rather than the much more susceptible 'Spanish' types (Lim, 1971).

Yellow Spot (Tomato Spotted-Wilt Virus)

Yellow spot of pineapple occurs in all production areas of the world, with the exception of peninsular Malaysia (Lim, 1985). Infection by the tomato spotted-wilt virus always kills the pineapple plant. Therefore, vegetative propagation does not transmit the virus to subsequent plantings (Kohrbach and Apt, 1986). Pineapple yellow-spot disease, caused by the tomato spotted-wilt virus, was first observed in Hawaii as a distinct disease in 1926 (Illingworlh, 1931). During the next 4 years, the disease spread and became a serious problem, with considerable rotting of fruits in the field. The initial symptom is a slightly raised yellowish spot, with a darkened centre, on the upper surface of the leaf. Shortly after formation of the initial spot, a characteristic chain of secondary spots develop and progress into a basal leaf and stem rot. Frequently, particularly on young plants, the rotting and cessation of growth on one side of the stem cause the plant to bend severely, eventually killing the entire plant. The disease can occur on the developing crown, with the rot progressing into the fruit and frequently causing distortion of the fruit (Illingworth, 1931). Infection occurs most frequently on pineapple plants during early growth. On occasion, crowns on developing fruit may also become infected (Linford, 1943). The transmission of the yellow-spot virus by the onion thrips, *Thrips tabaci* (Lindeman), was shown in 1932 by Linford. Pineapple is not a preferred host of thrips, but the thrips still move into pineapple fields from adjacent weed hosts and probe the plants, leaving the virus in the tissue (Linford, 1932). Later, the yellow-spot virus was shown to be identical to the tomato spotted-wilt virus (Sakimura, 1940).

Black Rot

Black rot, also called *Thielaviopsis* fruit rot, water blister, soft rot or water rot, is caused by the fungus *C. paradoxa* (De Seynes) Sacc. (syn. *T. paradoxa* (De Seyn.) Hohn (telemorph *C. paradoxa* (Dade) C. Moreau). The disease is a universal fresh-fruit problem but normally not a problem with processed fruit, because times from harvest to

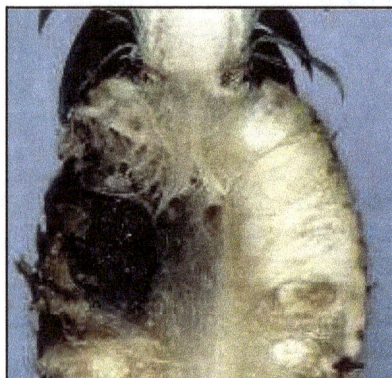

Figure 20: Black Rot

processing are too short for infection. The severity of the problem is dependent on the degree of bruising or wounding during harvesting and packing, the level of inoculum on the fruit and the storage temperature during transportation and marketing. Black rot does not occur in the field unless fruit is overripe or injured. Black rot of the pineapple fruit is characterized by a soft water' rot, which usually starts at the point of detachment of the fruit. Diseased tissue turns dark in the later stages of the disease because of the dark-coloured mycelium and chlamydospores. Infection of the pineapple fruit occurs through wounds resulting from harvesting and postharvest handling. Susceptibility varies, with the 'Red Spanish' types being more resistant than 'Smooth Cayenne'. Under conditions of high humidity, conidia may readily be produced on pineapple residue and be disseminated by wind to the unharvested fruit. Inoculum levels on fruit at harvest vary according to the environmental conditions prior to harvest (Rohrbach and Schmitt, 1994). The high correlation between moisture (rainfall duration) prior to harvest and disease following harvest has resulted in the name 'water rot'. Infection occurs within 8-12 h following wounding. Refrigeration at 9°C during transportation will slow development of the disease, but, when fruit are returned to ambient temperatures, disease development will resume (Rohrbach and Phillips, 1990). Dipping of fruits for 5 min in Thiabendazole 1000 ppm or Benomyl 3000 ppm would minimize rotting. Avoiding injury to fruit during harvest and transit will prevent disease occurrence.

Fruitlet Core Rots (Black Spot)

FCR (Rohrbach and Apt, 1986) or black spot (Keetch, 1977) (also called fruitlet brown rot and eye rot (Snowdon,

1990)) is a descriptive term for a brown to black colour of the central part of an individual fruitlet. FCR is caused by an infection by a pathogen or, more commonly, a group of pathogens. Botanically the central area of the fruitlet core is the septa (inverted Y) between the three seed cavities or locules. Because individual or mixtures of pathogens may be associated with the FCR symptom, there is considerable confusion in the literature. In addition to the multiple pathogens two mites have also been reported to be associated with the occurrence of FCR epidemics (Rohrbach and *Apt*, 1986. It is theorized that the very low levels are the result of botanical malformations of individual fruitlets caused by disruptions in the normal phyllotaxis of the fruit (Kerns *et al.*, 19361. Malformation of the fruitlet allows infection of the stylar canals and nectary ducts by a range of pathogens. In contrast, true epidemics result from the coincidence of optimum environmental conditions resulting in predisposed flowers, production of inoculum of the pathogen(s) and transport of the inoculum to potential infection sites.

Each major pineapple production area appears to have characteristic pathogens associated with the FCR symptom, probably as a result of the environmental conditions of the area (Rohrbach, 1980). For example, in Hawaii, *Penicillium* and *Fusarium* species are most commonly associated with FCR (Rohrbach and Apt, 1986). In South Africa, *Penicillium* species are most commonly found (Keetch, 1977), while, in Brazil, *Fusarium* species are most commonly associated with the FCR symptom (Bolkan *et al.*, 1979).

FCR and associated symptoms are of major economic significance only as epidemics, not at endemic levels. Fortunately, true epidemics are relatively sporadic in the

'Smooth Cayenne' cultivar and in the major commercial pineapple areas of the world. The disease could become more important if some of the more susceptible, low-acid 'Smooth Cayenne' cultivars and hybrid cultivars are grown commercially for fresh-fruit markets.

Fusariosis

Fusariosis is caused by the fungus *F. subglutinans* which is the conidial stage of G. *fujikuroi* Edwards. The disease, first described in Argentina in 1954, was first reported in Brazil in 1964 and within 10 years had spread over the entire country (Laville, 1980; Rohrbach, 1983). The fruit symptoms at low severity levels are similar to those of FCR, which vary from light through medium to dark brown, extending partially to completely down the fruitlet core. FCR from *Fusarium* sp. is usually a 'dry' type of rot). In Brazil, the symptom is not limited to a single infected fruitlet, as in typical FCR reported in other pineapple production areas. Fruit symptoms involve multiple fruitlets, with the infected area of the fruit surface appearing off-colour initially and later becoming sunken, with profuse pink sporulation and exudation of gum (Figure 21). Gum exudation can be confused with the exudation from *Thecla* wounds (Laville, 1980).

In Brazil, the disease causes major losses in the three major cultivars, Terola', 'Jupi' and 'Smooth Cayenne' (Rohrbach, 1983), Levels of fruit infection can vary from 5 to 75 per cent (Laville, 1980). Infection is thought to occur through open flowers, although major levels of disease also occur from inoculations to the developing inflorescence (Ventura *et al.*, 1981). Infection of the inflorescence and fruit also occurs from injuries caused by insects, particularly

Figure 20: Pink Disease

the bud moth,. Once the developing fruit is infected, secondary infections can occur on the developing slips or suckers. The infected seed material is then distributed to new planting areas, thus infesting new sites. Soils can remain infested for several months. Spread within infested fields is primarily by insects but may also be by wind (Laville, 1980). Control of fusariosis is most effective by planting disease-free seed material and by controlling insects, particularly the bud moth (Laville, 1980). Fungicides, such as captafol at 700 g a.i. ha $^{-1}$, starting at differentiation through harvest at 20-day intervals, have given good control of the fruit-rot phase in Brazil (Bolkan *et al.*, 1978).

Pink Fruit

Pink disease can be found at very low levels in most pineapple production areas of the world (Rohrbach, 1983).

However, economicaly significant epidemics are only known to occur in Hawaii, the Philippines and Taiwan. When epidemics occur in Hawaii and Taiwan, the highest incidences occur in February, March and April. In the Philippines, however, epidemics occur from August to September (Hine, 1976). Pink disease of pineapple fruit is characterized by the typical symptom of brown to black discoloration of the fruit tissue when heated during the canning process. Depending on the bacterial strain and the severity of the disease, symptoms in uncooked fruit may be completely absent or may include extremely severe fruit translucence, light pinkish to brownish colour of the fruit cylinder, and/or a 'cantaloupe-like' odour (Rohrbach and Apt, 1986). *E. herbicola* in uncooked pineapple fruit is essentially symptomless and very difficult to detect. *G. oxydans* in uncooked fruit induces pinkish brown to dark brown discolorations and may have a 'cantaloupe-like' odour. *A. aceti*–more recently classified as *Acetobacter liquefaciens* (Gossele and Swings, 1986) in uncooked fruit with only a few fruitlets infected can be symptomless. However, in moderately to severely infected fruit (many fruitlets infected), symptoms range from pinkish brown to dark brown (Rohrbach and Pfeiffer, 1976a; Kontaxis and Hayward, 1978). The symptom is reported to be caused by the bacteria producing 2,5-diketogluconic acid, which reacts with amino acids to form brown to black pigments (Buddenhagen and Dull, 1967). Strains of the acetic acid bacteria, such as *A. liquefaciens,* have been reported to produce browning and rotting of apples and pears and to have been isolated from guava, mango and Surinam cherry (Gossele and Swings, 1986).

In contrast to the other fruit diseases, the economic significance of pink disease is the inability to detect diseased fruit prior to processing, with the result of brown to black slices in a sealed can. Thus, quality control during processing is critical to detection of low levels and management of diseased fruit in the cannery (Rohrbach and Apt, 1986). In fresh-fruit production, low levels of pink disease are not of major economic importance. However, when high incidences occur, with strains having symptoms prior to cooking, economic loss can occur.

Marbled Fruit

Marbling disease is caused by the acetic acid bacteria *A. peroxydans* Visser 't Hooft and *E. herbicola* van *ananas* (Serrano) Dye. Marbling disease has been reported in essentially all pineapple production areas of the world (Rohrbach, 1983). However, epidemic levels occur only in the lowland tropics, where temperatures remain above 21-27°C during fruit development (Rohrbach and Apt, 1986). In production areas such as Thailand, where disease incidence is high, from 5 to 20 per cent of the slices in the cannery will be marbled. In Hawaii, highest levels occur in April and May although the disease may occur at any time. Low fruit acid and brix are also associated with high levels of the disease. The fruit disease of pineapple termed 'marbling' is represented in the literature by a wide range of symptoms. The most common symptom is a yellowish to reddish brown to very dark dull brown discoloration of the infected fruit tissue. Infected tissues generally become hardened, granular and brittle in texture, with colour variation in the form of speckling (Rohrbach and Apt, 1986). The disease may affect individual fruitlets but more typically affects a group of fruitlets or the entire fruit. Frequently, the speckled appearance will occur in vascular tissues in the core of the fruit. The diseases reported as bacterial fruitlet brown rot (Serrano, 1928) and fruitlet black rot (Barker, 1926) have generally been considered to be a variation of marbling. An additional symptom, termed 'brown and grey rot', has also been associated with marbling disease. The causal organism of brown and grey rot will also cause marbling symptoms. In general, much less is known about marbling disease than about pink disease. Currently, no known controls of marbling exist. When epidemics occur, infected fruit can be detected and excluded prior to going through the cannery by external appearance and a test to measure fruit firmness, such as sticking a knife into the fruit. If incidences are extremely high, all fruit must be discarded. In contrast to pink disease, marbled-fruit tissues can be discarded before being packing in cans. Currently, no known controls of marbling exist. When epidemics occur, infected fruit can be detected and excluded prior to going through the cannery by external appearance and a test to measure fruit firmness, such as sticking a knife into the fruit. If incidences are extremely high, all fruit must be discarded. In contrast to pink disease, infected fruit tissues that are not discarded before processing can be discarded before packing in cans.

Internal Browning

Internal browning, also termed endogenous browning or black heart, is a physiological disorder of pineapple fruit. The disorder is of major significance in Australia, Taiwan, Kenya and South Africa, where fruit are grown and harvested at or near frost conditions (0-10°C). The disorder is also very important in the marketing of fresh fruit when refrigeration is used to extend shelf-life. Internal browning is of economic importance only where fruit are grown under very cool conditions or are refrigerated for long periods through marketing channels prior to consumption (Paull and Rohrbach, 1985). Internal browning is characterized initially by a small greyish translucent zone beginning at the base of the fruitlet adjacent to the fruit core. This zone later darkens, becoming brown to black. When symptoms are severe, the entire internal fruit tissues are brown to black, thus giving rise to the name 'black heart'. No organisms have ever been shown to be associated with the internal browning symptom. Internal browning is thought to occur from increased polyphenol oxidase activity (Teisson, 1979; Paull and Rohrbach, 1985). Low ascorbic acid levels have been associated with symptom expression (Paull and Rohrbach, 1985). Symptoms may develop in fruit that has matured in the field at low temperatures in the range 5-10°C. Symptoms may also develop within 4 days of ambient temperatures following refrigeration at common commercial shipping temperatures of 7°C (Rohrbach and Apt, 1986).

Pink Disease

Pink disease is of little importance in fresh fruit, but can be a very serious sporadic problem in processed fruit because of the lack of detection prior to canning. It was first reported in Hawaii by Lyon (1915) and is now known in Australia, the Philippines, South Africa and Taiwan (Rohrbach, 1983). Cultivars vary in their susceptibility (Rohrbach and Pfeiffer, 1975). At least three genera of bacteria have been reported to cause pink disease: *Erwinia,*

Gluconobacter and *Acetobacter* (Rohrbach, 1976a; Kontaxis and Hayward, 1978). The *Erwinia herbicola* species has recently been redescribed as *Pantoea citrea*, based on an isolate from the Philippines (Cha *et al.*, 1997). Species of the remaining genera are the acetic acid bacteria *Gluconobacter oxydans* and *Acetobacter aceti* (Cho *et al.*, 1980). Pink-disease bacteria are vectored to the pineapple flowers by insects and mites, probably attracted to the nectar. Honey-bees may play a role as important vectors of *Gluconobacter* and a lesser role for *Acetobacter* (Gossele and Swings, 1986). The nectar is thought to provide a nutrient source for the survival of the bacteria until they become latent in the nectary gland or stylar canal or locule. Gossele and Swings (1986) suggest that the bacteria may actually overwinter in honey-bee hives. Once the bacteria are inside the flower, they remain latent until the fruit matures, sugar concentrations increase and translucence occurs. Pink disease has been controlled with insecticides, which are thought to control insect vectors. Where pink disease occurs sporadically, insecticide applications have not been economic. In the Philippines, where pink-disease epidemics are seasonally predictable, pink disease has been controlled with applications of disulphoton. Disulphoton at 0.83 kg a.i. ha $^{-1}$ per application starting at the red-bud stage and with three additional applications at 5-day intervals (throughout flowering), has resulted in the highest level of control (Kontaxis, 1978). Applications of ethephon to inhibit flower opening and reduce nectar flow have resulted in partial but significant control of fruit collapse (Lim and Lowings, 1979a) and pink disease. Forcing plantings, so that flowering does not coincide with fruiting in adjacent plantings, which may have fruit collapse, can reduce disease development. Applications of Bordeaux mixture have resulted in variable control (Lim, 1985). Any treatment that reduces bacterial heart rot may reduce primary inoculum levels for flowering plants. Sanitation is an important factor in initial low incidences of fruit collapse. Infected plants and fruit should be destroyed or removed from the field, as they may provide a source for secondary inoculum.

Fruit Collapse

Bacterial fruit collapse, caused by *E. chrysanthemi* Burkh. *et al.*, is only economically important in peninsular Malaysia, although the bacteria have been reported elsewhere on pineapple (Melo *et al.*, 1974; Chinchilla *et al.*, 1979; Rohrbach, 1983). The disease is thought to be indigenous to Malaysia (Lim and Lowings, 1979b). The economic importance of fruit collapse in Malaysia is probably due to the use of the much more susceptible 'Singapore Spanish' cultivar (Lim, 1985). Symptoms of fruit collapse usually appear on maturing fruit 2-3 weeks prior to normal ripening. Infected fruit are characterized by exudation of juice and release of gas, as evidenced by bubbles. Fruit shell colour becomes olive-green. Dissection of completely infected fruit shows only cavities within the skeletal fibres of the fruit (Lim, 1985). Bacterial fruit collapse is caused by *E. chrysanthemi* Burkh. *et al.*, and the initial inoculum comes from other infected fruit. Insects such as ants, beetles and flies, are vectors of the bacteria, transporting them to flowers from other collapsed fruit or from plants with bacterial heart rot. Ants, as well as other insects, are thought to be attracted to the nectar available there. Open flowers are the infection site where the bacteria enter the developing fruit. The bacteria remain latent in the ovary until 2-3 weeks before normal ripening when sugar levels begin to increase rapidly and polyphenoloxidase levels decline (Lim and Lowings, 1978).

Management of Pests and Diseases on Seed Material

Pest and disease-free seed materials are critical to preventing the establishment of insects and pathogens in newly planted pineapple fields. The presence of mealybugs, scales and mites, as well as *Fusarium* infected seed materials, must be monitored at the seed source before transport for planting, in order to implement effective controls. The pineapple red mite will only become a problem on stored seed under dry conditions. Mealybugs, scales and the red mite can be controlled by dipping seed in an approved insecticide, such as diazinon (Petty and Webster, 1979). Red mites can also be controlled by orientating seed material in its normal vertical position, so that the leaf axils collect natural rainfall or dew, or by methyl bromide fumigation of the seed material (Osburn, 1945). The blister mite can be controlled by dipping seed materials in an approved miticide, such as endosulphan. Fusarium infected seed has been hot-water-treated at 54°C for 90 min with benomyl at 50 g 100 I^{-1}, but growth was retarded and up to 50 per cent of the plants were killed (Maffia, 1980).

Plant and Fruit Abnormalities

In addition to pests and diseases, some fruit and plant abnormalities are found to occur, which make plants less productive and fruits useless. The abnormalities and their probable causes are as follows.

Crown without Fruit

A rare case was noticed where peduncle elongated and produced a crown without producing inflorescence. The crown developed normally, but not the fruit. Two reasons are attributed to it: (1) lack of food reserves for developing fruit as the plant was small and (2) an imbalance in the growth hormones (George and Oomen, 1968).

Multiple Crowns

Multiple crowns are a common disorder that can be of genetic or environmental origin. Multiple crowns increase the size of the fruit core and result in flattening of the upper portion of the fruit, which reduces the value of the fruit for the fresh market and for canning. Collins (1960) states that fasciation–an abnormal growth resulting in two to numerous crowns–is relatively uncommon in 'Smooth Cayenne' and 'Queen/ but common, in 'Singapore Spanish' and Ternambuco'. Collins (1960) also reported the existence of mutant clones of 'Smooth Cayenne' that produced 50 per cent or more fruit with multiple crowns. Environmental conditions that promote the multiple-crown disorder are high fertility nnd rapid growth following a period of prolonged drought, if such conditions occur about the time of inflorescence initiation (Collins, 1960; Py *et al.,* 1987). An increased incidence of multiple crowns is correlated with periods of high irradiance and high temperature that occur during early inflorescence development, although the disorder is generally thought to be due to high temperature injury. Increased planting density reduced the incidence of multiple crowns (Norman, 1977; Scott, 1992), presumably because mutual shading in the more dense plantings reduced the temperature of the reproductive apex. The significance of this shading effect is confirmed by the observation that there was a higher incidence of deformed crowns on outside rows than in the interior of a field.

Irrigation during inflorescence development reduced the incidence of multiple crowns (Py et al., 1987).

Ordinarily fruit bears a single crown but in some cases fruit bears more than one, even to the extent of 25. Consequently the top of the fruit will be flat and broad and fruit will be unfit for canning. Such fruits taste insipid and are corky. It is supposed to be a heritable character, found mostly in Cayenne group (George and Oomen, 1968). Multiple crowns are more frequent with low planting densities (Linford and Mehrlich, 1934) and when floral initiation occurs during dry-and-sunny period (Py *et al.*, 1987). In the latter case, irrigation reduces number of fruits affected. Vigorous growth with abundant fertilizer or planting on virgin ground, encourages multiple crowns (Linford and Spiegelberg, 1933).

Fruit and Crown Fasciation

Fasciated fruits are deformed to such an extent that they are totally useless. In certain cases, proliferation is so

Figure 22: Fruit and Crown Fascinated Fruits

extreme that fruit is highly flattened and twisted with innumerable crowns, as high as 173 on it, stretched over a length of 76 cm in a twisted plane. Fruits and crowns fasciation is associated with high vigour of plants. Such plants take longer time to flower. High fertility of soil and warm weather, where conditions are highly congenial for vigorous vegetative growth, may favour fasciation (George and Oomen, 1968). Excessively high temperatures during floral differentiation and calcium or zinc deficiency are said to be some of the causes for fasciating inflorescence (Py *et al.*, 1987). The incidence of fasciation was found to increase with advancing ratoons (Chadha *et al.*, 1977). The fasciation is due to some kind of accident affecting normal control of growth sequence during ontogeny (Collins, 1960). In addition, prevalence of favourable climate for vegetative growth during flower differentiation is supposed to cause this abnormality.

Collar of Slips

The collar of slips is typified by the presence of a large number of slips arising from stem close to the base of the fruit, or even directly from the fruits itself. The excessive slip growth is at the expense of the fruit resulting in small, tapered fruits, often with knobs at the base. High nitrogen fertilization (Ganapathy *et al.*, 1977; Gonzales Tejera and Gandia Diaz, 1976) and high rainfall along with relatively low temperature are supposed to be congenial for such an abnormality (Ganapathy *et al.*, 1977).

Variations from true collar of slips type such as 'Near collar' and 'Knobby fruit' also occur.

Long Tom

The 'Long tom' is distinguished by excessive length and small diameter of fruit, which is usually knobby at the base and matures late. Sucker development is slow and slips are numerous.

Dry Fruit and Bottle Neck

The dry fruit and bottle neck fruit types are very similar and may be derived from the same parent. In dry fruit type, fruit is small, flowers are absent and fruitlets do not develop. In bottle neck, lower fruitlets develop normally and upper ones do not develop and give the same appearance as dry fruits. Suckers are freely produced from both the types.

Figure 23: Bottle k in Pineapple Fruits

Sun-Scald

This results when plant leans or falls over to one side, thus exposing one side of the fruit to direct sunlight. The cells of the exposed surface get damaged. Later shell surface assumes a brownish to black colour and cracks may appear between fruitlets. In high-density planting, intensity of sun-scald is very much minimized. Under favourable climates where leaf growth is luxuriant leaves can be tied around the fruits to protect them from sun-scald. The other method is to cover sun-exposed portion of the fruit with dry straw or grass or with any other locally available materials.

In addition to these abnormalities, branching of the peduncle and proliferation of leaves and fruits have also been observed.

Miscellaneous Fruit Diseases, Pests and Deformities

Numerous pineapple fruit diseases and blemishes of minor occurrence and importance have been noted and described, much of it in unpublished form. The minor importance of these symptoms and abnormalities results from their sporadic occurrence and lack of economic effects on pineapple production. Thus, little has been done to determine the cause and aetiology of these symptoms and abnormalities.

Yeasty Fermentation

Yeasty fermentation is caused by the yeast species *Hanseniaspora valbyeusis* (M. Okimolo, unpublished data) and can be a major problem in overripe fruits. A dry yeast rot has been attributed to *Candida intermedia* var. *alcohololophila* (M. Okimoto, unpublished data). Occasionally, the disease will occur in green fruit, having severe interfruitlet corking symptoms with associated fruit cracking. The disease has also been associated with high incidences of fruit sunburn (Lim, 1985). Losses can be minimized by reducing sunburn and harvesting fruit before they are overripe.

Glassy Spoilage

Glassy spoilage is caused by infections with the yeast *C. guilliermondii* and may be associated with fruitlet core rots (M. Okimoto, unpublished data) caused by mixed infections of yeast and *Penicillium* or *Fusarium* species. As with yeasty fermentation, losses can be minimized by harvesting before fruit are overripe.

Acetic Souring

Acetic souring is caused by bacteria and is characterized by an offensive odour similar to that of a mixture of organic acids, including acetic acid. Juice of infected fruit may be very viscous and cloudy with bacteria. No controls are known.

Miscellaneous Fruit Rots

Fruit rots caused by *Aspergillus flavus, Botryodiplodia theobromae* and *Rhizopus oryzae* or *Rhizopus stolonifer* have been reported as postharvest diseases (Snowdon, 1990). A fruit rot caused by *Hendersonula toruloidea* (Natt.) has been reported by Lim (1985) on the Mauritus cultivar. Green fruit rot caused by *Phytophthora* species occasionally causes large losses of lodged first-ratoon fruit in Australia under very wet conditions. These pathogens generally require some form of wounding for infection. Commercially, these diseases are of very minor importance.

Miscellaneous Physiological Fruit Diseases

Woody Fruit

Woody fruit is a disease of unknown cause. The disease is characterized by brown streaks distributed throughout the fruit tissue, which is very woody and hard in consistency. The disease is associated with certain clones of 'Smooth Cayenne' and therefore is assumed to be of genetic origin. Roguing at harvest is used to eliminate seed material from plants showing symptoms.

Sun Scald and Frost Injury

Frost injury occurs occasionally in some areas in Australia. Symptoms are shell discoloration and cracking between the eyes. Multiple crowns (two or more) develop when young fruit are exposed to high temperatures early in the development stage. Because nothing is known about the stage of fruit development most susceptible to high temperatures, no controls are available. The problem is mainly important where fruit are to be sold fresh with the crowns attached. In Australia, multiple crowns are trimmed to one to improve appearance and to facilitate packing in boxes. Fasciation is an abnormal development of the inflorescence and crown, resulting in a flattening of the upper part of the fruit with multiple crowns ranging from two to many. Fasciation has been associated with genetic and environmental conditions, although Py (1952) has indicated that the phenomenon is not hereditary. Cultivars such as the 'Smooth Cayenne' are less susceptible than the 'Singapore Spanish'. Within the 'Smooth Cayenne' cultivar, certain, clones are much more susceptible than others.

Miscellaneous Fruit Pests

Trephritid Fruit Flies

Prior to 1953 methyl bromide was used to treat pineapple imported into continental USA. Since then, it has been demonstrated that 'Smooth Cayenne' pineapple cultivars (low- and high-acid types, with at least 50 per cent 'Smooth Cayenne' parentage) are not hosts for the tephritid fruit flies: Mediterranean fruit fly, *Ceratitis capitata* (Wiedermann), melon fly, *Dacus cucurbitae* (Coquillet), and the oriental fruit fly, *Dacus dorsalis* Hendel (Bartholomew and Paull. 1986).

Management of Postharvest Diseases and Pests

Black rot is commercially managed by minimizing bruising of fruit during harvest and handling, by refrigeration and with chemicals. Fruit must be dipped in an appropriate fungicide within 6-12 h following harvest prior to packing and shipping (Rohrbach Phillips, 1990). Internal-browning symptom development can be reduced by waxing with paraffin-polyethylene waxes at wax-to-water ratios of 1:4-9 (Rohrbach and Apt, 1986). Waxing has been shown to increase internal CO_2 concentrations, thereby lowering O_2 concentrations, which results in reduced polyphenol oxidase (Paull and Rohrbach, 1985).

The *Penicellium induced* FCR, LP and IFC fruit diseases have been reduced by applications of endosulphan (3.35 kg a,i. ha^{-1} in 2338 1 water) at forcing and 3 weeks following forcing. Reductions have been significant but only under low to moderate disease levels (Le Grice and Marr, 1970; Rohrbach *et al.*, 1981; Rohrbach and Apt, 1986). Control of typical FCR induced by *F. subglutinans* has not been demonstrated. Control of fusariosis is most effective by planting disease-free seed material and by controlling insects, particularly the bud moth (Laville, 1980). Hot-water treatment of seed material at 54-C for 90 min with benornyl at 50 g 100 1^{-1} is effective for disinfestation but will retard growth and kill up to 50 per cent of the plants (Maffia, 1980). Fungicides, such as captafol, at 700 g a.i. ha^{-1}, starting at differentiation through harvest at 20-day intervals, have given good control of the fruit-rot phase in Brazil (Bolkan *et al.*, 1978). Resistance to fusariosis occurs in *Ananas* and *Pseudoananas* (Laville, 1980).

Scale can be controlled relatively easily by preharvest applications of an appropriate registered insecticide, taking into consideration last application to harvest residue restrictions.

Integrated Pest Management

Environmental and food-safety concerns have focused attention on IPM. The concept of IPM is to employ several techniques simultaneously to solve specific pest and disease problems for the long term rather than in the short term. Success relies on an in-depth understanding of the pineapple production system and the ecology and biology of each pest or disease and associated organisms (*e.g.* vectors, natural enemies). Emphasis must be placed on the importance of each pest or disease from an economic, biological and ecological perspective. In order to evaluate the importance of the pest or disease, efficient techniques are needed to monitor changes in populations of pest and levels of diseases or pathogen populations. The changes must be correlated with yields and quality.

In most pineapple production systems throughout the world, mealybug wilt must be controlled by the management of ants and mealybugs. Severe infestation may have an impact on the production system and the final product in several ways. As a direct pest, feeding reduces plant growth, fruit quality and yield. The presence of mealybugs on fresh fruit may become a quarantine issue, as well as a quality issue when present in the canned product. The indirect effect and the most severe impact are the resulting mealybug wilt, with high rates of field infestation. Ants play a major role in the impact of mealybugs and mealybug wilt on pineapple. Soil tillage during fallow eliminates essentially all in-field ants, and new infestations must move into the newly planted field from field border areas. The rate of establishment of permanent ant colonies and mealybug wilt is relatively slow (Beardsley *et al.*, 1982). When ants are controlled, the parasitoid *Anagyrus ananatis* Gahan and other biological control agents can maintain populations of the pink mealybug below damage thresholds in Hawaii. The use of Amdro® is efficacious, allowing natural biological control agents to function while reducing overall insecticide usage.

Techniques for monitoring ants, using trap stakes, have been developed (Beardsley *el al.*, 1982). Recommendations are to use trap stakes baited with peanut butter/soybean oil at intervals of 30m (100 ft) along field borders of new plantings. Trapping must be done in late afternoon, with data being taken after darkness occurs. The first monitoring should start at 3 months following planting and be repeated at 3-month intervals thereafter. When mils are detected, they may be controlled with site-specific applications of ant baits or insecticides. Other monitoring techniques, such as pit-fall traps, honey-vial traps and pineapple-juice traps, have also been used. Threshold levels of ants have not been very well defined and the presence of any ants is considered problematic. Populations of mealybugs have been monitored with sticky tape placed in the lower part of the pineapple plant. Relatively high levels of mealybug are required for mealybug wilt. Diagnosis of mealybug wilt virus-infected seed material can be done rapidly and inexpensively, using a tissue-blot immunoassay system (Hu *et al.,* 1993), in order to establish virus-free pointings. Reniform and root-knot nematode threshold levels at planting for pineapple production in Hawaii have not been well defined soil, provides excellent control. However, stored seed results in poor uniformity of early plant growth and can reduce crop yields. Planting fresh planting material results in more rapid uniform growth. Freshly removed seed material for immediate planting must be dipped in a fungicide within 12 h of removal. Currently, seed materials are dipped in triadimefon. Black rot is commercially controlled in fresh fruit by minimizing bruising of fruit during harvest and handling, by refrigeration and with chemicals. Fruit must be dipped in a fungicide within 6-12 h following harvest prior to packing and shipping. Currently fruit can be dipped in triadimefon. The 'Red Spanish' cultivar is generally more resistant to *C. paradoxa* than 'Smooth' Cayenne', but, due to low yields and poor quality, this is not an economically viable cultivar.

Pineapple scale only becomes a problem when the balance of biological control is upset, for example by the application of residual, broad-spectrum insecticides (Sakirnura, 1966). Quantitative monitoring techniques have not been developed. While a monitoring technique has been developed for the tar-sonemid mite, correlation with the penicillium-induced fruit diseases has not been well enough established to predict disease.

IPM for pineapple production systems has met with varying success and has not been broadly implemented for several reasons. First, less expensive alternatives are still available. The annual application of Amdro® for ant control in pineapple is much less expensive than the labour required for a detailed ongoing monitoring programme. As long as other alternatives are available, farmers will not learn and implement monitoring activities. Second, in Hawaii, the importation and development of biocontrols have essentially reached a standstill because of environmental concerns for non-target species. Until agriculture is forced by economic or regulatory incentives to implement IPM, traditional pest <ind disease strategies will be used. Thirdly, IPM does not generally reduce pest and disease levels low enough to meet quarantine requirements, thus requiring other pest- and disease-control strategies. In Hawaii, an IPM verification programme has been established, which was modelled after the national IPM protocol for potatoes. Multidisciplinary teams, including members from industry, research and extension, identify pests and diseases and recommend IPM practices. IPM protocols are developed based on establishing the best management approaches. Verification of producer practices is done by farm visits and review of records, in order to assign points in relation to each IPM protocol. High scores allow producers to use IPM as a marketing tool and to better educate farmers and consumers as to the value of products grown under IPM principles.

Chapter 10
Harvesting and Yield

Harvesting

Pineapple flowers 10-12 months after planting and attains harvesting stage 15-18 months after planting, depending upon the variety, time planting, type and size of plant-material used and prevailing temperature during the fruit development. Irregular flowering results in harvesting spread over a longer period. In nature, pineapple comes to harvest during May to August. This results in prolonged vegetative phase; and supplies of fruits to factories cannot be properly regulated. Beside fruits which mature in winter are acidic. There is also a scope of altering fruit size and maturation with the use of chemicals or plant regulators. The stage of harvesting in pineapple is very important. If it harvested at immature stage, it does not develop its full sugar content and flavour. If left until it is too ripe it loses its flavour and appearance resulting in flat-

and-insipid fruits. Hence, it is very essential to harvest fruits at an ideal stage of ripening. In Kew pineapple, Mookerji *et al.* (1969) found harvesting at 137 days after flowering as optimum Thrissur (Kerala), India for canning purposes. Studies on optimal harvest maturity of Kew pineapple in North Bengal have shown that fruit harvested between 115 and 130 days after flowering were better suited for canning (cited by Bose, 1985). In Bangalore (India), Chadha *et al.* (197 noticed increase in fruit size up to 165 days after flowering. Biswas *et al.* (1979) also observed close correlation between fruit weight and days taken from flowering to harvest. The disparity in these observations might be because of high temperatures prevailing at Thrissur and North Bengi as a result of which required summation of heat units is achieved in shorter period than at Bangalore.

The most commonly followed index of harvesting Kew pineapple yellowing of the basal 1/2 to 2/3rd portion of the fruit for local marking and yellowing of basal 1/3 to 1/2 portion for distant market and canning factories. Harvesting should be done with a sharp knife, severing fruit stalk with a clean cut, retaining 3-5 cm long stalk attached to fruit Crowns are not detached. Though, maturity or harvesting standards a prescribed, experience is necessary to decide the exact degree of ripeness for harvest.

Fruits are packed in baskets woven with bamboo strips. For local markets, they are arranged in baskets lined with paddy-straw to stand on their stumps. The second layer of fruits is arranged on the crowns of the first layer of fruits. Each basket weighs 20-25 kg. For distant markets, fruits are wrapped individually with paddy straw and then packed.

Ratooning

The common practice is to renew plantation once in 4-5 years. The practice of continuing plantation for 20-30 years is seen in hill-side planting of Assam (Chadha, 1977). Experiments conducted at the Indian Institute of Horticultural Research, Bangalore, on ratooning in the high-density (55,500 plants/ha) plantings have revealed that the average fruit weight in the first and second ratoons was 88 per cent and 79 per cent of the plant-crop; and the plant stand also reduced, leading to reduction of fruit yield by 49.3 per cent and 46.2 per cent in the first and second ratoon crops (Chadha *et al.*, 1977). Prolonged ratooning would also result in reduction of flowering plants, consumer appeal of the fruit and size, and the number of fruits suitable for canning, but increase fasciated fruits. It was not possible to prevent reduction of fruit yield in ratoon crops by increasing irrigation or by higher doses of nitrogenous fertilizers (Rao *et al.*, 1977).

Yield

Yield of pineapple varies with variety, agroclimate, agrotechniques followed, the type of planting material used and the planting density. Rao (1946) estimated pineapple yield at 6 tonnes/ha in India. The yields in Assam were 22 tonnes/ha (Chowdhury, 1947); but Bhattacharya and Sarma (1949) had put this figure at 12-15 tonnes/ha. The average yield in south India, irrespective of the variety, was 14 tonnes/ha and with Kew variety it was 25 tonnes/ha (Naik, 1963). According to the estimates of the Indian Institute of Foreign Trade, the average yield in India for pineapple is 10-15 tonnes/ha. It was consistently demonstrated at the Indian Institute of Horticultural

Research and other institutes engaged in pineapple research in India that with the adoption of high-density planting and improved agrotechniques, high yields of pineapple, comparable to yields at Hawaii, the Philippines, Australia, Malaysia and Taiwan, can be obtained. Yield of pineapple could be forecasted based on the plant-growth characteristics. Number of leaves one year after planting and number of suckers in a plant could be used as indices of the size of the fruit with the following prediction equations (Chadha *et al.,* 1977)

1. Fruit weight in kg = 1,3987 + (0.0028 x No. of leaves)

2. Fruit weight in kg = 1,238 + (0.2520 x No. of suckers)

Chapter 11
Postharvest Management

Fruit Physiology and Development
Fruit Development

Bract, calyx and ovary tissues have become fused within and between fruitlets during development to form the collective fruit. No floral abscission occurs, so the withered style, stamens and petals can be found on mature fruitlets. The large bract subtending each fruitlet is fleshy and widened at its base and bends over the flattened calyx surface, covering half of the fruitlets. Cell division is completed prior to anthesis and all further development is the result of cell enlargement (Okimoto 1948). Fruit development studies (Sideris and Krauss, 1938) have shown the fruit weight and its components (core, fruitlets, the collection flesh, fruit shell) increase in a continuous sigmoid fashion once the inflorescence has been initiated. Fruit mass increases about 20-fold from the time of flowering until maturation (Singleton, 1965, Teisson and Pineau,1982).The number of fruitlets comprising a fruit varies widely with plant condition and environmental conditions. A typical smooth cayenne fruit has about 150 fruitlets, which produce a mature fruit weighing about 202 kg (Tay, 1977). Fruit dry- matter content can vary with the conditions prevailing during fruit development.While the crown probably has no direct effect on the growth of the fruit (Senenayake and Gunasena 1975; Chen, 1999), crown growth increases for about 30-45 days after fruit growth has commenced. Crown removal early in fruiting does not always lead to greater fruit weight. Crown size can be reduced by the plant growth regulator chloroflurenol when applied at he flowers stage, naphthalene acetic acid (NAA) and 2-(3- chlorophenoxy)-prop ionic acid (3-CPA) have also been used to reduce crown size (Bartholomew and Criley, 1983). Preliminary work also suggests that the crown may play a role in fruit translucency development (Paul and Reyaes 1996). Translucency is when the flesh has a water soaked appearance.

Crown size is an aesthetic character of economic concern for packing and is generally part of the grading standard. There does not appear to be a relation throughout the year between crown, fruit size and stem starch the crown photo assimilates seem to be delivered from its own photosynthesis.

Under short day lengths and cool night temperatures, natural induction of inflorescence developments occurs. This procedures flowering disrupts harvest scheduling, increases the number of passes necessary to harvest a field and, if small plants are induced, leads to smaller fruit. Applying plant growth regulators is a possible approach to limiting environmental induction or transgenic plants with reduced environmental sensitivity may be produced.

Fruit Physiology

The half yellow stage is regarded as ripe and is near the maximum fruit weight if still on the plant (Wardlaw, 1937). The Most marked changes in flesh composition occur in the 3-7 weeks prior to the half yellow shell-colors stage. Just prior to the half yellow stage, fruit translucency can start to develop, with this development continuing after harvest.

Senescence related loss of membrane integrity leads to water soaked, translucent flesh, which tends to be softer than non–translucent fruit. Gortner and Leeper (1969) have shown that amides nitriles, simple esters and salts of phenoxy acids, naphthalene compounds, phenyl acid and trichlorophenoxyacetic acids can delay fruit senescence. NAA and phenoxyacetic acid are the most effective in delaying shell yellowing when applied as brief postharvest dips. These chemicals induce little change in total soluble solids (TSS), acidity, carotenoid pigments or vitamin C content; both chemicals are, however, phytotoxic to the crown. There are no marked changes in fruit texture during ripening. Water loss can lead to some reduction in fruit firmness.

Shell Chlorophyll

Chlorophyll levels show little change until the final 10-15 days before full shell yellowing and then decline. Shell carotenoid pigments remain reasonably constant during the phase, declining only slightly before rising again as the fruit senesces. Flesh carotenoids increase during the final 10 days before the full ripe stage. A similar decrease in shell chlorophyll and an increase in carotenoid occur in harvested fruit.

Respiration

The non climacteric pineapple fruit produces around 22 ml kg^{-1} h^{-1} of carbon dioxide at 23°c, with no dramatic respiratory change during ripening. Ethylene production increases during ripening, but has no pronounced peak. Exogenous ethylene application stimulates only respiration rate when there is some chlorophyll remaining in the shell, and may also open the crown leaf stomata. The absence of a peak in ethylene production and lack of a relationship of respiration with pronounced bio chemical ripening changes support the conclusion of a non-climacteric pattern of development.

Organic Acid Metabolism

Juice PH declines from 3.9 to 3.7 as fruit approach the full yellow stage and increase as the fruit senesces with titratable acidity showing the opposite trend. The flesh acidity increases distally from the central core (4 mEq 100 ml $^{-1)}$ outwards to (4 mEq 100 ml $^{-1)}$, and a major portion (65-70 per cent) of the total non volatile acids occurs as free organic acids. The two major non-volatile organic acids are citric acids and malic. Malic acid can vary from 18 per cent to 30 per cent of total acids in pineapple and does not vary markedly between the cool and warm season crops, but can fluctuate threefold with weather conditions that favour water evaporation. Malic acid accumulates when sunlight and evapotranspiration are low. Citric acid (28-66 per cent of total acids) is lower in smooth cayenne fruit harvested in the warm season and tends to vary primarily with stage of fruit development. Citric acid increase uniformly with fruit development peaking before malic acid and before full ripeness. Fruit malic acid levels do not change after harvest or during and after storage. During Storage at 7°C, citric acid increases about 25 per cent and then declines slightly when fruit are held at 22.5°C.The citric acid content of unstored fruit does not change though there is a slight decline in titratable acidity after harvest.

Ascorbic acid content varies significantly with the clone and increase with increasing solar radiation and air temperature. Ascorbic acid is correlated with a clone acidity does not contribute substantially to titratable acidity and is 25 per cent higher near the surface of the fruit than near the core. Ascorbic acid levels can vary from 200 mg 1^{-1} in smooth cayenne to 710 mg 1^{-1} in spiny Guatemala. The level of fruit ascorbic acid at harvest has been negatively related to the intensity of internal browning symptoms associated with postharvest chilling injury (Combres, 1979; Paul and Rohrbach, 1982). Internal browning is a minor problem if fruit ascorbic acid content is greater than 500 um.

Sugar Metabolism

Sugar content plays an important role in the flavors characteristics and commercial assessment of pineapple fruit quality. TSS, mainly sugars, are soften used as an indicator of fruit maturity and quality. TSS can vary by 40g 1^{-1} from the more mature, sweeter basal tissue to the crown end of the fruit, and decline only slightly after harvest. Starch is not accumulated as the fruit ripens, though it is high during early fruit growth which could explain the absence of dramatic changes in TSS postharvest.

The major sugars in mature fruit are sucrose, glucose and fructose and the peak in sucrose concentration is attained at full yellow stage and then declines. Fruit sugars continued to increase through to senescence, unless the fruit is harvested. Chen (1999) showed that total soluble sugar content is low during fruit growth and composed mainly of glucose and fructose. Glucose is at a slightly higher concentration than fructose during the early stage of fruit development. Sucrose accumulated rapidly 6 weeks before commercial harvest and ultimately exceeds the glucose and fructose concentration. Fructose and glucose continue to increase postharvest.In addition sucrose accumulated more in the fruit let than in the interfruitlet tissue until the last 2 weeks of fruit development, when sucrose accumulation rate in the interfruitlet tissue was greater than in the fruit-let.

Three sugar metabolic enzymes (sucrose synthase, sucrose phosphate synthase and invertase) are thoughts to control sugar accumulation by fruit tissue. The activity of sucrose synthase is high in younger pineapple fruit and declines to a very low level 6 weeks before harvest, while the activity of sucrose phosphate synthase is relatively low and constant throughout fruit development. The activities of acid, neutral and cell wall invertase are high in the younger fruit and decline to low levels 6 weeks before harvest, when sucrose starts to accumulate. The activity of cell wall invertase does increase 4 weeks before harvest, mainly in fruitlet tissue, while the activities of neutral invertase (NI) and acid invertase (AI) remain low, concomitant with the accumulation of sucrose, indicating that these enzymes may be a prerequisite for sucrose accumulation in pineapple fruit flesh. The high activity of

CWI, favoring apoplastic phloem unloading, may play a role in sugar accumulation in pineapple fruit flesh at the later stages of fruit development.

Fruit Cell Walls

The cells walls of pineapple fruit parenchyma are regarded as unlignified, but do contain esterified ferulic acid. Overall, the non-cellulosic walls are intermediate between the unlignified poaceae ant typical dicotyledon cell walls. Glucuronoarabinoxylans are major component of the non cellulose fraction. Xyloglucans, along with smaller amounts of pectin polysaccharides and glucomannas, are present. Pineapple juice contains pre-dominantly galatomannans and no acid sugars are detected, suggesting little pectin hydrolysis during ripening. This juice neutral polysaccharide forms a gum on processing equipments, which is readily hydrolyzed by commercial cellulose, hemicellulase and pectinase preparations. Glucan $(1{\to}3$, $1{\to}4)$ linkages are absent, in contrast to poaceae cell walls.Glucuroarabinoxylans have also been isolated from lignified cell walls of pine apple leaf.

Volatiles

A wide range of volatiles (157 compounds) have been identified, including esters, lactones, aldehydes, ketones, alcohols and a group of miscellaneous compounds. esters constitute over 80 per cent of total volatiles. Free and glycosidically bound constituents have been found including 2- pentanol,2-butoxyethanol,hexanoic acid,phenol,4-hydroxybenzaldehyde,vanillin and syringaldehyde, as a glycons.Volatile esters increase both on the plant and after harvest.The volatiles vary with cultivar and are higher in summer fruit, especially ethyl alcohol and ethyl acetate.

Translucent fruits are higher in ester content, such as ethyl acetate which is very low in opaque fruit. Information on the temperature of individual components to the fruits aroma and flavors is still lacking, though furaneol and ethyl-2-methylbutanoate are the two largest odour contributors.

Lipids and Amino Acids

Total lipids decline at maturity, as does squalene, while phopholipid, total sterol and amino acid values increases (Selvaray *et al.*, 1975). Free amino acids in the juice are at the minimum at the middle of the fruit growth (Gortner and Singleton, 1965). The exception is free methionine, found at low levels until the onset of ripening, when it increases to 0.7 mM at senescence.

Phenols

The phenolic acids (p-coumaric, ferulic and sinapic) have been tentatively identified in pineapple fruit. These acids, except sinapic, are in fruit showing internal browning ans are the substratre for polyphenol oxidase activity (Van Lelyveld and de Bruyn, 1977).

Enzymes

Peroxidase activity falls steadily during fruit development (Gortner and Singleton, 1965), reaching a minimum of one third the initial value during ripening. Acid peroxidase (pH optimum 5.0) does not appear to be related to chilling injury symptom development (Teisson, 1977), though it does not decline during storage (Teisson and Martin–Prevel, 1979).

No ascorbic acid oxidase was detected in one report (Van Lelyveld and de Bruyn, 1977): however, ammonium sulphate–precipitated protein has been shown to have ascorbic acid oxidase that has a high pH 6.0 (Teisson, 1977). The PPO has optimum pH near 5.0 and a temperature optimum near 45 C (Teisson, 1977). Das *et al.* (1997) found three PPO isomers, with the major isoform being a tetramer of identical subunits (c. 25 kDa) and having an optimum activity between pH 6 and 7. Acetone powders of pineapple stem extracts contain the proteinase bromelain and a family of poly peptide inhibitors of this enzyme (Reddy *et al.*, 1975). Proteinase activity appears abruptly after flowering, remains high during fruit development (Gortner and Singleton, 1965; Lodh *et al.*, 1972). The major Systeine proteinase is fruit bromelain (EC 3.4.22.33), a non glycosylated proteinase that is immuunologically distinct (EC3.4.22.32) (Rowan *et al.*, 1990). Fruit bromelein has an estimated molecular weight of between 23 and 31 kDa by different laboratories (Yamada *et al.*, 1976; Ota *et al.*, 1985). The fruit bromelein is an acidic protein and its isoelectiric point (pI)is 4.6, different from the basic stem bromelein, the pI of which is 9.6 (Yamada *et al.*, 1976). The amino terminal end of the fruit bromelain has an additional alanine (Yamada *et al.*, 1976). Fruit bromelain has a 20 per cent cross reactivity with the anti stem bromelain (Sasaki *et al.*, 1973).

Maturity, Harvesting and Handling

Quality Criteria

A definition of pineapple fruit quality often refers to the sum of those characteristics of a fruit that makes it more palatable and therefore desirable to the consumers. Quality criteria used have been appearance (size, condition, shape), colour (shaell and flesh), taste (sugars, acids), aroma, flesh translucency, texture and fibre content. Skin colour is the most common measure of physiological maturity and consumer expectations of quality (Dull 1971). For more consumers, food value and vitamin content are also considerations for fruit quality purposes sugars are estimated as TSS by refractometry as percentage (g 100 ml^{-1} juice) or as brix. Problems soon arise in applying these criteria: TSS to acidity ratio is often used as a measure of flavour, though acid variation has a greater impact on the ratio than on TSS. A 20 per cent of TSS with 1 per cent acidity will not taste the same as the 10 per cent TSS with 0.5 per cent acidity. In apineapple the problem of TSS and acidity is compounded by the effect of cultivars, season, maturity, stage at harvest position in the fruit and fruit development conditions. The ascorbic acid content in smooth cyenne varies from 130 mg100 ml^{-1} juice at the base of the fruit to 280 100 ml^{-1} juice at the less mature top of the

Figure 24: Differenent Stages of Maturity

Figure 25: Ripened Fruit

fruit. A similar variation occurs for titrable acidity (TA) (Singleton, 1955). TSS shows the opposite trend, having 19 per cent TSS at the base and 15 per cent at the top. While the TSS /acidity ratio may be constant after maturity; both acidity and TSS decline with the increasing shell colour from about the 30 per cent yellow shell colour stage (Tay 1977). Seasonal variation in sugar and acid value can be significant. Acid levels are higher during cool season and sugars (TSS) are lower. The reverse occurs in warm season. Commercially TSS/acid ratio is regarded as the most reliable measure of fruit flavour. It was recommended that only pineapple with total suspended solids (TSS) of 14°Brix and above be marketed as fresh fruit.

Harvesting

Pineapple of hand harvested, with pickers being directed as to stage of shell colour required. Picking early in the morning and protection of harvested fruit from the sun can reduce the heat load at cooling. Fruits are generally selected, brokened from the peduncle by pickers and placed on the conveyer belt running on a boom to transfer the fruit to truck field bins or field packed. Field packed fruits show significantly less fruit bruising and crown injury. When the field bin with fruit arrives at the packing shed fruit may be unloaded witby hands, by submerging the field bin in water. Fruits with more translucencies (Sinkers) are separated at this stage. The sinkers are highly translucent fruit that are more fragile and have poor storage life. High translucency is also associated with bacteria and yeast fermentation and acetic souring during handling, shipping and marketing. The water in the dump tank needs to be chlorinated and replaced frequently to prevent build up of disesase organisms. Care is taken to avoid damage to the crown leaves.

Composition

Freshly harvested pineapple fruit contains 80-85 per cent water. 12-15 per cent sugars, 0.6 per cent acids, 0.4 per cent proteins, 0.5 per cent ash, 0.1 per cent fats, some fiber and vitamins (mainly A and C). The vitamin C content varies from 10 to 25 mg/100 g. The range of chemical constituents of ripe pineapple depends on the stage of fruit ripeness, and on agronomic and environmental factors (Gortner *et al.*, 1967). The major carbohydrate constituents in pineapple fruit are the simple sugars sucrose, glucose, and fructose. There is no starch accumulation in pineapple fruit (Gortner *et al.*, 1967). The major acids in pineapple are citric and malic. Unlike citric acid, malic acid content changes dramatically with changes in environmental conditions. Although the fruit contains substantial ascorbic acid this does not contribute substantially to fruit acidity. The ripened fruit contains higher levels of glycine, alanine. methionine, and leucine, whereas lysine, proline, histidine, and arginine. are present at relatively low levels (Cortner *et al.*, 1967). Chlorophyll and carotenoids are the major pigments in green and yellow pineapple fruits. Several volatile compounds were identified (Dull 1975) in canned pineapple juice. These include acetic acid, 5-hydroxymethylfurfural, fomlaldehyde, acetaldehyde, and acetene (Table 15). Recently Katsumi *et al.* (1992) reported volatile constituents of green and ripened pineapple. Among a total of 157 constituents identified, 50 were identified for the first time in pineapple. The esters ethyl acetate and ethyl-3-(methylthio) propionate constituted over 80 per cent of total volatiles (Table 16). In ripened pineapple, ethyl acetate and butane-2-3-diol diacetate were the main constituents.

Table 15: Nutritive Value of Ripe Fruit Flesh in Pineapple. Composition of pineapple fruits.

Constiuent	Per cent (Fresh weight basis)	Constiuent	Per cent (Fresh weight basis)
Brix	10.8-17.5	Pigments (ppm of carotene)	0.2-2.5
Sucrose	5.9-120	Carotene (mg)	0.13-029
Glucose	1.0-3.2	Xanthophyll (mg)	0.03
Fructose	0.6-2.3	Esters (ppm)	0.2-2.5
Cellulose	0.43-0.54	Vitamins fresh wt	17-22
Pectin	0.06-0.16	Aminobenzoic acid	2.5-4.8
Titratable acid (as citric acid)	0.6-1.62	Folic acid	200-280
Citric acid	0.32-1.22	Niacin	75-163
Malic acid	0.1-0.47	Pantothenic acid	69-125
Oxalic acid	0.005	Thiamine	20-88
Ash	0.30-0.42	Riboflavin	10-140
Water	81.2-86.2	Vitamin B6	0.02-0.04
Fiber	0.30-0.61	Vitamin A	10-25
Nitrogen	0.045-0 115	Ascorbic acid	10.8-17.5
Ether extract	0.2		

Postharvest Fungicides and Waxing

In countries where pineapple is grown on extensive level fruits are commercially treated with fungicides in a dip or spray applicatuion to control postharvest fruit rot, caused by the fungus Chalara paradoxa. A wax frequently containg polyethylene/paraffin-based. May also be applied to the fruit along with fungicides. The major adventage of

Table 16: Volatile Compounds of Pineapple

Compound	Pineapple Flesh (ug/kg)
Acetoxy acetone	9[b]
γ-Butyrolactone	2-4[b]
γ-Caprolactone	120[a]
Chavicol	270[a]
2,5-Dimethyl-4-hydroxy-3(2'H) furanone	1200[a]
Dimethyl malonate	60[b]
Ethyl-3-accetox yhexanoate	6[b]
Ethyl-3-hydroxy hexanoate	3[b]
Ethyl-3-methyl thiopropionate	90[b]
Methyl-3-accetoxy hexanoate	30[b]
Methyl-3-hydroxy butvrate	6[b]
Methyl-3-hydroxy hexanoate	21 [b]
Methyl-3–methyl thiopropionate	120[b]
Methyl-cis-octenoate	0.9[b]
γ-Octalactone	300[b]
λ-Octalactone	300[b]
trans-Tetrahydro-3.4.5-tr methyl-5-vinyl furfuryl alcohol	0.9[b]

waxing is the reduction of the internal browning symptoms of chilling injury. Waxing also reduces postharvest water loss and improves fruit appearances. If the wax injures the cron only the fruit body is waxed.

Fruit Grading

The fruit to be packed needs to be mature, firm, well formed, free of defects, with flat eyes and a minimum of 12-14 per cent TSS. Fruits are graded based upon recognized appearance characteristics: degree of skin colouration, size (weight) absence of defects and diseases and other market needs before packing. Crown size is the crucial grade component, with a minimum size and a ratio of crown to fruit length of 0.33-1.5 for the higher grades. Crowns developed during the summer tend to be larger and may require gouging at harvest to meet the standard. This gouging leaves a wound for possible pathogen entry and degrades over all appearance. Gouging two months before harvest, avoiding visible scarring, is also practiced to limit crown growth (Soler, 1992 a).

Packing

Fruits are packed normaly in to cartons of two different sizes. A large carton (20 kg) containing 10-16 fruits for surface or air shipment with 5-6 fruits. Tourist packs of 2-4 are also prepared. Absorbant pads are used at the bottom of the carton and between the layers if fruits are alternately within the carton. In other packs fruits are placed vertically.

Export Grading and Packing

For reduction of postharvest disease incidence, the fruit should be treated, by dipping or spraying, with a solution of Dowicide A (sodium 2-phenylphenolate) at a concentration of 7 g per litre of water. Size grading and packing should be carried out immediately after treatment.

Pineapples are packed according to the stage of ripeness and the size of the fruit. Fruits in each cartons should be the same size, resulting in a range of counts. Accepted counts are as follows:

The preferred method of packing is to place the fruit vertically on the base, and then to place dividers between the fruits to prevent rubbing and movement. With some cartons, this is not possible and fruits are laid horizontally in alternating directions; where two layers of fruit are packed, a layer of card is required between the layers.

☆ 6 count–1.75 kg fruit (3.8 lb)

☆ 12 count–1.25 kg fruit (2.7 lb)

☆ 12 count–1.00 kg fruit (2.2 lb)

☆ 20 count–0.75 kg fruit (1.6 lb)

Fruits are normally packed to a net weight of 10 to 15 kg (22 to 33 lb) depending on the carton and the market. High value small pineapples may be shipped in some instances at 6 kg (13 lb), whereas the large fruit in some cases may be packed up to 20 kg (45 lb).

A full-telescopic two-piece fibreboard carton with internal dividers between the fruit; bursting strength 275 lb/in 2. Top and bottom ventilation, in addition to side vents are required, particularly where sea-shipments in break bulk are used. Where staples are used in carton construction, care should be taken to ensure complete staples closure to prevent fruit damage.

Storage

The problem with fresh pineapple is not how to store the fruit so that it can be ripened for use by the consumer, but rather how best to store the fruit in order to minimize loss of quality in the fruit at the time of harvest. Adequate ventilation is required to ensure arrival of fruits in perfect condition at the retail outlet if the fruits are to be transported for a shorter distance. Care should be taken to prevent bruising during harvesting and packing, and finally fruits have to be adequately protected against fungal infection.

When fruits are to be transported for long distances or for a period of several days, refrigerated transport is required to slow down ripening process. In tropical areas, pineapples could be stored well for 20 days when refrigerated at 10-13°C, provided fruits used for storage are healthy and unbruised (Huet and Tisseau, 1959). Fruits harvested in initial stages of ripeness remained saleable for longer after storage at 45°F. The total shelf-life of mature green- and half-ripe fruits was longer than those of ripe fruits. Akamine and Goo (1971) opined that 8°C is the optimum temperature for Smooth Cayenne fruits harvested at a relatively advanced stage of ripeness. The level of atmospheric oxygen in transport container can be reduced to slow down respiration. By maintaining oxygen at 2 per cent by controlled nitrogen enrichment, Akamine and Goo (1971) succeeded in prolonging life of the fruits from 1 to 3 days. The best storage is at 7.2°C and 80 or 90 per cent RH. The fruits of Red Spanish cultivar so stored for 7 days remained marketable for 9 more days than when they were kept at 15.6° C and 75 per cent RH (Cancel, 1974).

Within the cold storage area, humidity should be maintained nearing to saturation level and air should be replaced wherever possible. However, at lower temperatures with longer storage period, there is a marked increase in acidity. In Smooth Cayenne, there was a 35 per cent increase in acidity in fruits stored at 8°C for 10 days (Tisseau *et al.*, 1981). Storage at a temperature of less than 7°C may lead to serious deterioration of tissues. When temperature was around 20°C, there was a marked drop in acidity over a few days with subsequent breaking down of tissues (Tisseau, 1982), starting from epidermal and sub epidermal tissues.

Storage life of Kew pineapples at 70°F both in summer and winter was found 12 days, but by coating fruits with wax emulsion, it increased to 16 days (Bose *et al.*, 1962), and by treating fruits with 500 ppm 2,4, 5-T and then coating with wax emulsion and storing them at 70°F increased their storage life to 30 days. Fruits could be retained in good marketable condition up to 41 days at the Bidhan Chandra Krishi Vishwa Vidyalaya with the treatment of fruits with NAA and GA$_3$ at 500 ppm and 100 ppm. The untreated ones stored well only for 12-15 days at room temperature. GA$_3$ delayed development of yellow colour and NAA enhanced it. In addition, NAA was effective in preventing storage rot. In another study, mature and ripe pineapples were stored for 4 weeks at 11-13°C and 8-9°C and 85-90 per cent relative humidity. Experiments carried out at the Hisar Agricultural University, indicated possibility of extending storage life with Benlate treatment (Bose, 1985).

A. Low Temperature Storage

As for many other fruits and vegetables, refrigeration aids shipping and extends storage life of pineapples. One half-ripe fruit (smooth cayenne) can be held for about 2 weeks at 8.5-12.5°C, gaining an additional 1 week of shelf life (Dull 1975). Mature green fruits (fruit picked with no yellow color showing on the shell) are susceptible to chill injury when stored at temperatures less than 10°C (Chadha *et al.*, 1974). Py *et al.* (1974) recommended a storage temperature of 8.5°C for South African pineapple. Endogenous brown spot frequently occurs simultaneously with chill injury. This is not always the case, however, since storage at 8°C for 1 week, followed by 1 week at 21°C, frequently induces endogenous brown spot but not chilling injury (Gortner 1967). About 1 week of additional storage life of pineapple can be gained for each 6°C decrease in storage temperature for fruit showing about 25 per cent shell yellowing at harvest. According to Dull (1975) at 7°C, the maximum storage life was about 4 weeks. A practical method to control the endogenous brown spot of pineapple has been discovered: dry heat is applied at 35°C for 24 h. either before or immediately after cooling (Akamine 1963). Pineapple fruits with and without crown can be stored for 20 days at low temperature (10°C) without much loss in physiological weight. On the other hand the fruits can also be stored for 7-10 days at room/ambient temperature (20°C) without effecting their physico chemical and quality characteristics. (Singh and Attri 1999)

B. Controlled-Atmosphere Storage

The effects of different levels of carbon dioxide and oxygen on respiration and storage behavior of pineapples have been reopened. Stage 4 fruits (Akamine 1963) were placed in atmospheres containing air with 21, 10, 5, and 2.5 per cent oxygen. Nitrogen made up the balance of the atmosphere. The respiration rate decreased with decrease in oxygen concentration of the atmosphere. When stage 4 fruits were placed in an atmosphere of 21 per cent oxygen plus 0.03 per cent air, 5 and 10 per cent carbon dioxide, the rate of respiration was suppressed slightly due to the increase in the concentration of carbon dioxide (Akamine 1963). Dull (1975) concluded that decreased oxygen and increased carbon dioxide concentrations had no obvious effect on fruit quality. No major advantage of quality maintenance appears to be gained by manipulation of the concentration of these two environmental gases.

C. Postharvest Treatment

1. *Growth Regulators*

A postharvest dip of half-ripe smooth cayenne fruit in a 100-ppm solution of 2,4.5-trichlorophenoxyacetic acid (2, 4, 5-T) extended the shelf life from 6 to 14 days when the fruit was stored at ambient temperatures (Bourke 1976). The quality of the treated fruit at the end of the 14th day was judged to be good, while that of the control was fair to poor. Based on quality indices such as sugar. acid, pigment, and flavor, Gonner *et al.* (1967) concluded that the hormone functioned as a senescene inhibitor. Mature green Kew pineapples treated with 500 ppm of 2.4,5- T extended the shelf life of fruit by 12-30 days at 21°C (Das). A single application of NAA solution (50-200 ppm) increased flowering percentage, fruit size, and weight and prolonged time from flower differentiation to ripening by about 15 days (Das 1964).

2. *Other Chemicals*

The postharvest fruit rot of pineapple caused by *Thielaviposis paradoxa* can be prevented by dipping the pineapple stalk in benzoic acid or Shirlan. A fungicide more promising than Shirlan is imazalil. Infection was reduced by application of 1 per cent sodium salicylanilide within 5 h of cutting to the cut surface of the stem piece left on the fruit. Dowicide A is used commercially in Hawaii, but benomyl (0.63 lb active ingredient/acre) gave excellent control under experimental conditions (Cook 1975). The postharvest storage life of pineapples could be extended up to 2 or 3 weeks at 11.I°C and I week at room temperature when the fruits were dipped in coating material, drained. and packed into well-ventilated plastic canons (Ewing *et al.*, 1980). The effects of the type of surfactant. pH, and filmfonning materials were investigated. The treatment reduced the endogenous brown spotting of pineapples.

D. Irradiation

Preliminary studies on the effect of gamma radiation on the shelf life of fresh pineapple, conducted at the Hawaii Agricultural Experiment Station. University of Hawaii, indicated that irradiation with a 50 ha^{-1} dose extends the shelf life of pineapple, as judged by the delay in degreening of the fruit (Upadhya *et al.*, 1966), without impairing the physical and chemical or organoleptic characteristics of the fruit (Moy *et al.*, 1967-68), An irradiation dose of 30-500 laad was found to induce formation of cytotoxic substances in pineapples (Upadhya *et al.*, 1966-67) These substances disappeared in fruit irradiated at 30-50 had after 8 days of storage at 17.8°, but they persisted in fruit irradiated at 100 or 500 ha^{-1}. Upadhya *et al.* (1967-68) reponed that, like untreated fruit (Akamine 1963), storage conditions of 7.2-12.8°C and about 90 per cent RH were optimum for irradiated pineapples to extend their storage life. Pablo *et al.* (1975) concluded that low-dose irradiation may be a feasible method of extending the shelf life of fresh pineapples. which can tolerate a dose of about 50 had (Brewbaker 1964-65).

Postharvest Diseases and Disorders

Physiological and Physical Disorders

Chilling injury

Exposure of pineapples to temperatures below 7°C results in chilling injury. Ripe fruits are less susceptible than unripe or partially-ripe fruits. Symptoms include dull green color when ripened (failure to ripen properly), water-soaked flesh, darkening of the core tissue, increased susceptibility to decay, and wilting and discoloration of crown leaves.

Black Heart

Is usually associated with exposure of pineapples before or after harvest to chilling temperatures, *e.g.* below 7°C for one week or longer. Symptoms are water-soaked, brown areas that begin as spots in the core area and enlarge to make the entire center brown in severe cases. Waxing is effective in reducing chilling injury symptoms. A heat treatment at 35°C for one day has been shown to ameliorate EBS symptoms in pineapples transported at 7°C by inhibiting activity of polyphenol oxidase and consequently tissue browning.

Injury/Losses during Transportation

Pathological Disorders

Thielaviopsis Rot (Black rot, Water blister)

Caused by *Thielaviopsis paradoxa*, may start at the stem and advance through most of the flesh with the only external symptom being slight skin darkening due to watersoaking of the skin over rotted portions of the flesh. As the flesh softens, the skin above readily breaks under slight pressure.

Yeast Fermentation

Caused by *Saccharomyces spp*, is usually associated with overripe fruit. The yeast enters the fruit through wounds. Fruit flesh becomes soft and bright yellow and is ruptured by large gas cavities.

Figure 26: Brown Spot

Figure 27: Water Blister

Figure 28: Fruitlet Core Rot

Control Strategies

1. Careful handling to minimize mechanical injuries
2. Prompt cooling and maintenance of optimum temperature and relative humidity throughout postharvest handling operations.

3. Application of fungicides, such as thiabendazole (TBZ).

Processing

Canned Products

Most of the world's pineapple production is canned. Slices are the most valuable product, followed by pineapple juice, chunks, and diced pineapple. Some other products are fruit salads, sugar syrup, alcohol, and citric acid. Canning is the most common preservation method used for pineapple. Pineapple from west coast has been found to be suitable for canning. In general, the fruits are soned by size, shape, and freedom from major blemishes. The varieties Smooth Cayenne, Giant Kew, Valeva, Balanca. and Espinola Roja were found to be suitable for canning (Bhat *et al.*, 1972; Bhat *et al.*, 1977). Ripe and juicy fruit with characteristic colour and aroma should be selected and washed in fresh water.

Remove the crown, peel eyes and the core of the fruits (cores removed with a corer) and cut into transverse slices of 1.25 mm thickness with a stainless steel knife. Prepare the sugar syrup by mixing 1 cup of sugar with 1 cup of water and bring it to boil. Place the pineapple rings (5-6) in butter size plain canes. Fill the inter space with syrup leaving 1.25 cm headspace.

Exhaust the cans until the temperature at the center of the can contents attains 80°C (time taken is about 7 minutes) and seal. Process the cans in boiling water for 20 and 25 minutes in the case of butter size and A2 ½ cans respectively. Cool the cans quickly and store.

Pineapple Juice

Pineapple juice contains neutral polysaccharides composed predominantly of galactomannan. Pineapple juice is usually prepared as a by-product in the canning industry and is a delicious beverage. Entire fruits or even scrapings and cores can be used for extraction of the juice. The juice can be consumed as such or with plain soda.

After washing, remove the crown of the fruits by giving a sharp twist: remove the peel with a stainless steel knife. With a sharp V-shaped stainless steel knife, remove the eyes. Discard the damaged portions. Cut the sound portions into small pieces: pass them through a mincer or chop them finely with a sharp stainless steel knife. Wrap the prepared fruit in thick cloth and press out the juice in a small basket press or worm type juice extractor. Strain the juice through course muslin cloth. The juice obtained is generally tart and becomes palatable after adding sugar. Add sugar according to taste (60 g/kg) and strain the juice again through course muslin cloth.

Heat the prepared juice rapidly in an aluminium or stainless steel pan on direct fire to a temperature of 82-85°C. Pour the hot juice into plain cans leaving 0.6 cm head space; seal the cans immediately. Alternatively pour the juice into previously sterilized, warm bottles to overflowing and seal air tight with crown corks.

Process the cans (milk size) in boiling water and the bottles in water at 80-82°C for 25 minutes at sea level. At higher altitudes increase the processing time by 2 minutes for every 300 m rise in altitude. Cool the cans immediately in running water; allow the bottles to cool gradually. Store the cool products in a cool, dry place.

The degumming of juice can be achieved by using pectinase, cellulase, and hemicellulase preparations (Chenchin *et al.*, 1984). The retention of ascorbic acid in bottled and sulfited juice and squash was 80-85 per cent at room temperature (24-30°C) and only 38-47 per cent at 37°C (Pruthi and Lal 1954). Color deterioration was three times more rapid at 37°C than at room temperature. Canned juice can be stored for 12-15 months without any serious loss in quality or nutritive value (Pruthi and Lal 1954).

Pineapple Squash

Put the pineapple peel and core in a nonreactive saucepan and add enough water to cover them. Bring to a boil, then simmer for 30 minutes. Strain the liquid through a cloth and press the juice out of the peel and core. Measure the mixture, and for each cup (1/4 litre) use 1 cup of sugar. Heat the juice with the sugar over low heat until the sugar is dissolved. Cool the mixture. Add the lemon juice. Pour into clean bottles, cork and store.

Pineapple Jam

After removing the crown, peel and the eyes, cut the sound portions to the fruit into small pieces. Crush them thoroughly and obtain a uniform mass. Add an equal quantity of sugary by weight to the prepared fruit; allow it to stand for ½ -1 hour. Cook the mixture slowly with vigorous stirring till the temperature reaches 105.5°C (at sea level) or till the mass approaches the jam consistency. Fill the jam into sterilized dry jars; cool them and screw tightly. Store in a cool, dry place.

Concentrate

Pineapple juice concentrate can be prepared by various methods, with or without neutralizing the excess acidity. It can be prepared under vacuum and fortified with ascorbic acid (Sandhu *et al.*, 1985). Aroma concentrate from pineapple juice can be separated and later added back to stripped juice concentrate (Remteke 1987) to obtain fully flavored concentrate. The concentrate with added aroma was comparable to fresh juice and was superior to all the other concentrates in sensory qualities (37). Pineapple juice concentrate could be stored satisfactorily at 10-38°C with the addition of SO_2 without causing any undesirable change in color or flavor (Sandhu *et al.*, 1985).

Powder

Attempts have been made to prepare powder from pineapple juice by freeze-drying (Phanindrakumar *et al.*, 1991). Giant Kew variety of pineapple with a minimum 12° Brix, acidity around 1.0 per cent, and with a minimum of 120 µg total carotenoids per 100 g juice was found to yield a product of acceptable quality.

Wine

For proper utilization of culled fruits of pineapple a technology for making a quality wine was developed by Attri and Singh 2002. They further found that Variety Queen had an edge over that of Kew for its over all quality and commercial acceptance.

Waste Products

Several products such as bromelain, wine, sugar syrup, citric acid, wax, sterols, and cattle feed can be produced from the mill juice of pineapple processing industry waste (Joseph and Mahadeviah, 1988).

Table 17: Chemical and Sensory Qualities of
Fruit and Fruit Juice

Quality	Variety	
	Giant Kew	*Queen*
Weight of individual fruit with crown (kg)	1.75-30	0.7-1.2
Length of fruit (cm)	20-25	12-15
Diameter of fruit (cm)	12.0-14.5	8.5-11.0
Weight of crown (g)	50-100	150-200
Percent juice yield	50-55	35-40
Brix	7-14	12-14
Acidity (per cent) as citric acid	0.9-1.1	0.6-0.9
Total carotene (mg/100 g juice)	70-250	300-400
Color of juice	Slight yellow	Deep yellow

Phanindrakumar *et al.*, 1991.

Other Products

Pineapple Bran

Approximately 35–40 per cent of the fruit weight is in the skin and ends, proper solid waste disposal becomes a serious consideration in the pineapple processing operations. The simplest solution that is currently followed by many processors is to return it to fields and use it as soil amendments. Because of its high moisture content, it easily ferments and presents storage problems. To reduce this moisture content, other operations macerate the skin and ends to produce a course pulp and press out the liquid it contains-mill juice. The resulting solid material called as wet pineapple bran is suitable for cattle feed. If dried to a moisture level of less than 12 per cent, the product is known as pineapple bran, which can be easily stored.

Table 18: Chemical Analysis of Aroma and Stripped Pineapple Juice Concentrate

Sample	Esters	Carbonyls	Alcohols	Oxygenated Terpenes	COD
	(μg/100 g)	(μg/100 g)	(μg/100 g)	(μg/100 g)	(μg/100 g)
Fresh feed juice	15.0	2.00	0.180	0.303	10.70
Aroma concentrate	11.9	1.58	0.140	0.240	8.60
Stripped juice concentrate	1.2	0.16	0.015	0.025	085
Stripped juice concentrate with aroma added back	120	1.60	0.150	0.270	940

Remteke 1987.

Mill Juice

This is the liquid pressed from macerated pineapple skins during pineapple bran production. Although it has a lower soluble solid content and higher acidity than pineapple juice has approximately the same composition but contains some undesirable flavour materials and heat darkening agents, due to the Millard reaction. Juice from other off grade sources (*e.g.* rotton and damaged fruit) may be added to wet pineapple bran to increase bran yield or may be used as a source of sugar in the production of alcohol and vinegar. When prepared under the same sanitary conditions as pineapple juice and if treated further to remove the off flavour and off colour components, mill juice has been successfully used as a sweetner in packing medium.

Dietary Fibre

Processed pineapple peels with 80 per cent total fibre content and 1.0 kJg^{-1}, as prepared by the institute de investigaciones para la industria Alimentaria, Havana, Cuba, were used as a source of dietery fibre at 8 per cent maximum moisture in a dried beverage mix (Larrauria *et al.*, 1994, 1995). A mild laxative effect is observed. Fibre from cores has been evaluated as a food grade filler for some food products.

Candied Pineapple

Candied pineapple is probably one of the most popular dried pineapple products. The product is generally prepared using pineapple chunks or pineapple slices segmented with the knife. The chunks are soaked for 5 hours in sugar syrup (about 60 per cent Brix) containing an antioxidant, such as ascorbic acid or a sulphiting agent, to maintain colour and

Figure 29: Pineapple Products

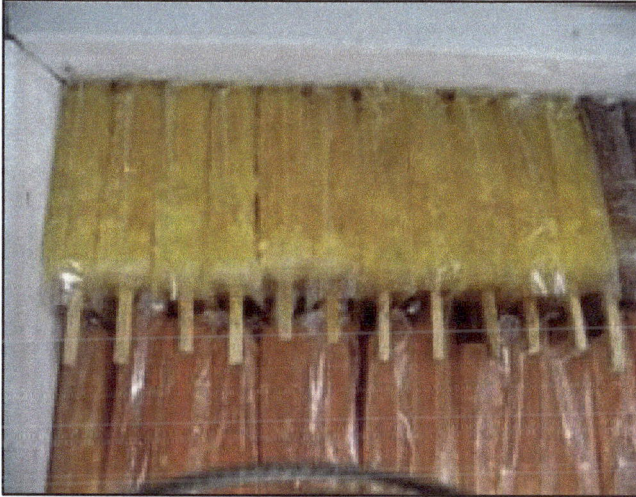

☆ Once you taste a pineapple, you'll never go for any other fruit (African proverb) (Swahili: Ukila nanasi, tunda lingine basi)

☆ Only the knife knows the heart of a pineapple.

are air dried. The dried pieces are sometimes coated with another type of syrup and dried again.

Ultra High Pressure Pasteurized Fresh Cut Pineapple

As a means of extending the shelf life, fresh cut pineapple chunks obtained from a commercial processor were packed in heat sealed polyethylene pouches and treated under various ultra high pressure (200, 270, and 340 MPa) temperature (Approximatelr 4, 21, 38°C) and time (5, 10, 40 and 60) min.) combinations (Aleman, *et al.*, 1994). Bacterial survival and total yeast and fungi counts generally decreased with an increase in processing pressure.

Prospects and Constraints

India has not been able to make a dent in the world market despite suitable climate and soils, large area under pineapple and a good processing infrastructure available in the southern region of the country. There are several reasons why Indian exports have not caught up over the years compared to other countries.

1. While the quality of Indian products has been meeting international standard, the price of these products has been high.

2. The productivity of pineapple crop needs to be improved for increasing production commensurable with area increase to bring down the cost of production to make them competitive in the world market.

3. As the cultivation of pineapple is highly capital intensive and is of semi-perennial nature, special credit facilities need to be provided.

4. Besides price competitiveness, the other constraints are lack of quality packaging, marking and labelling of Indian products.

5. India at present lacks an image in the export market, which is fully entrenched by several leading countries of the world. This is due to lack of publicity in the overseas markets.

6. While steps needs to be taken to improve performance of Indian pineapple processing industry through Governmental support mainly in the form of reducing cost of packaging, the major thrust should be in the direction of evolving an efficient distribution network within and outside the country. This calls for an integrated approach towards development of infrastructural facilities (transport and communications), development of

primary markets, improvements in packing, storage and handling facilities and techniques and subsidization of fresh produce movements.

7. Standardization of technology to bring down cost of production of fresh pineapple and its adoption by growers, assured market for the produce; and to produce pineapple throughout the year may go a long way in promoting pineapple industry in India.

References

Aghion, D and Beauchesne, G. 1960. Utilization de la technique de culture sterile d'organes pour obtenir des clones of d'*Ananas Fruits*, 15, 464-466.

AICFIP. 1978. *Annual Report of All India Coordinated Fruit Improvement Project* (Cell I), 1977.

AICFIP. 1982. Standardization of type and size of planting material in 'Kew' pineapple. Research reports of the All-India coordinated Fruit Improvement project (Cell-I), pp. 81-87.

Aich, K.1981. Ph. D. Thesis submitted to the Bidhan Chandra Krishi Vishwa Vidyalaya, Kalyani.

Akamine, E. K., 1976. Problems in shipping fresh Hawaiian tropical and subtropical fruits, *Acta Hort.* 57: 151.

Akamine, E.K., 1963. Fresh pineapple storage, *Hawaii Farm* sci. 12: 1-4.

Akamine, E.K., and Goo, T. 1971. Controlled atmosphere storage of fresh pineapple (*Ananas comosus* (L.) Merr.) Smooth Cayenne. *Res. Bull. Hawaii. Agri. Expt. Stat.* 152: 8.

Aldrich, W.W. and Nakasone, H.Y., 1975. Day versus night application.of calcium carbide for flower induction in pineapple. *Journal of the American society for Horticultural Science* 100,410-413.

Aldridge, N.L. 1960. Crookneckin pineapple is not hard to control. *Qd Agric. J.* 86:97.

Aldrigo, G. L. 1966. Pineapple nutrition. (In) : *Nutrition of Fruit Crops.* Ed. Childers, pp. 611-50.

Anderson, E. J. 1966. 1-3 Dichloropropene, 1-2 dichloropropane mixture found active against *Pythium arrhenomanes* in field soil. *Down to Earth* 22, 23.

Andre, E. 1889. Description et histoire des, Bromeliacell recoltees daus la colombie, I' Equateur et le vanezuela. G. Masson, Paris, 118 pp.

Anon. 1944. Pine tops, now wasted, may become important feed, experiment shows. *Hawaii Farm andHome, Honolulu Star Bulletin,* 7(12): 26.

Anon. 1975. DPI Digest–*the pineapple.* Department of Primary Industries, Queensland Government, Brisbane, 34pp

Anonymous. 2009. Statistical, Food and Agriculture Organization of United Nations.

Anonymous. 2010. Area and Production Statistics, National Horticultural Board Gurgoan, India.

Attri B.L.and D.B.Singh. 2002. Pineapple wine- an alternative to use culled fruits. *Indian Food Packer,* 56(4): 79-82.

Aubert, B. 1977. Etude du cycle de floraison naturelle de I' ananas 'Victoria' ala Reunion. *Fruits* 32, 25-41.

Aziz, T, Yuen, J.E. and Habte, M. 1990. Response of pineapple to mycorrhizal inoculation and fosetyl-Al treatment. *Communications in Soil Science and Plant Analysis,* 21: 19-20.

Baker, Kenneth and Colline, J.L. 1939. Notes on the distribution and ecology of Ananas and pseudananas in South America. *Amer. J. Bot.* 26: 397-702.

Balakrishnan, S., Aravindakshan, M. and Nair, N. K. 1978a. Efficacy of certain growth regulators in inducing flowering in pineapple (*Ananas comosus*). *Agri. Res. J. Kerala,* 16: 138-41.

Banziger, H. 1982. Fruit-piercing moths (Lep., Noctuidate) in Thailand: a general survey and some new perspective. *Mitteilungen der Schweizerischen Entomologischen Gesellchaft* 55, 213-240.

Barker, H. D. 1926. Fruitlet black rot disease of Pineapple. *Phytopathology* 16, 359-363.

Bartholomew, D.P. and Criley, R.A. 1983. Tropical fruit and beverages crops. In: Nickell, L.G. (ed.) *Plant Growth Regulating Chemicals.* CRC Press, Boca Raton, Florida, pp.1-34.

Bartholomew, D.P. and Kadzimin, S.B. 1977. Pineapple. In: Alvim, P.T. and Kozlowski, T.T. (eds). *Ecophysiology of Tropical Crops.* Academic Press, New York, pp. 113–156.

Bartholomew, D.P. and Malezieux, E. 1994. Pineapple. In: Schaffer, B. and Anderson, P. (eds) *Handbook of Environmental Physiology of Fruit Crops,* Vol.II. CRC Press, Boca Raton, Florida, pp. 243-291.

Bartholomew, D.P. and Paull, R. E. 1986. Pineapple fruit set and development. In: Monselise, S. P. (ed.) *Handbook of Fruit Set and Development.* CRC Press, Boca Raton, Florida, pp. 371-388.

Beardsley, J. W., Su, T. H., McEwen, F.l. and Gerling, D. 1982. Field interrelationships of the big-headed ant, the gray pineapple mealybug, and pineapple mealybug wilt disease in Hawaii. *Proceedings of the Hawaiian Entomological Society* 24, 51-67.

Benega, R., Cisneros, A, Hidalgo, M., Martinez, J., Arias, E., Arzola, M., Carvajal, C. And Isidron, M. 1998 b. Hybridization in pineapple results, and strategies to save time for obtaining and releasing new hybrid varieties for growers. *Abstracts of the third International Pineapple Symposium,* 17-20 November 1998. Department of Agriculture, Pattaya, Thailand, p. 16.

Bertoni, Moise. 1919. An Cient, Paraguay, *Series* II, 4: 250–322.

Bhat, A. V., G. Varkey, V. K. Sathyavathi. and J.S. Pruthi, 1977. Varietal trials in canning of pineapples. *Indian Food Packer* 3: 18.

Bhat. A. V., Y. K. Sathyavathi, V. K. George, arid K. Mookerjee. 1972. Effect of hydrochloric acid treatment on the crown and canning quality of pineapple fruit, *Indian Food Packer,* 26(6): 23.

Bhattacharya, S.C. and Sarma, K. 1949. Cultivation of pineapple (*Ananas sativus*) in Assam.

Bishop, E.J.B., Gradwell, J.B., Nell, J.A.G. and Bradfield, D.M. (1965) Pineapple plants: a good drought fodder for cattle. *Farming in South Africa*, 41: 6-7.

Black, CC. and Page, P.D. 1969. Pineapple growth and nutrition over a plant crop cycle in south eastern Queensland. II. Uptake and concentrations of N, P, K. *Queensland J. of Agric. and Animal Sci.*, 26: 385-405.

Black, R.F. 1962. *Qld. J. Sci.* 19, p.435.

Boher, B. 1974. Pineapple heart rot: histological study of infection by *Phytophthora palmivora:* Active penetration of the parasite into the aerial organs. *Fruits* 29, 721-726.

Bolkan, H. A., Dianse, J. C. and Cupertino, F. P. 1978. Chemical control of pineapple fruit rot caused by *Fusarium moniliforme var. subglutinans*. *Plant* Disease Reporter 62, 822-824.

Bolkan, H. A., Dianse, J. C. And Cupertino, F. P. 1979. Pineapple flowers as principle infection sites for *Fusarium moniliforme var. subglutinans*. *Plant Disease Reporter* 63, 655-657.

Boname. 1920. *Less ananas*. Paris.

Bose, T.K. 1985. Pineapple (by S.K. Sen). In: *Fruits of India: Tropical and Subtropical*. Naya Prokash, Calcutta.

Bourke, R.M. 1976. Seasonal influences on fruiting of rough leaf pineapples. *Papua New Guinea Agriculture Journal* 27, 103-106.

Braddock, R. J. and J. E. Marcy, 1985. Freeze concentration of pincapple juice. *J. Food Sci.*, 50(6): 1636.

Brewbaker, J.L., M.D. Upadhya and K.W. Chinf. Preliminary studies on the effect of gamma irradiation of pineapple. U.S. Atomic Energy Commission. Div. of Isotope Development. *Ann.* Rep. 1(64-65): 47.

Buddenhagen, I. W. and Dull, G. G. 1967. Pink disease of pineapple fruit caused by strains of acetic acid bacteria, *Phytopathology* 57: 806.

Burg, S.P. and Burg, E.A. 1966. The interaction between auxin and ethylene and its role in plant growth. *Proc. Natl. Acad. Sci. (U.S)* 55: 262-69.

Cabral, J.R.S, A.P. and da Cunha, G.A.P. 1993. Selection of pineapple cultivars resistant to fusariose. *Acta Horticulturae* 334, 53-58.

Camargo, L.M.P.C. and Camargo, O.B. 1974. Preliminary studies on inoculation techniques and on some aspects of the physiology of the fungus *Fusarium moniliforme V. Subglutinans*, the cause of pineapple gummosis. *Biologico* 40, 260-266.

Cancel, H.L., 1974. Harvesting and storage conditions for pineapple of the Red Spanish variety. *J. Univ. Puerto Rico* 58: 162-69.

Cann, H.J. 1961. *Pineapple growing*. Division of Horticulture, New South Wales, Department of Agriculture.

Cannon, R.C. 1952. Weed sprays in pineapples. *Queensland Agr. J.* 75: 139-41.

Cannon, R.C. 1960b. Recent development in weed control in pineapples. *Proc. 2nd Aust. Weeds Conf. Paper* 16: 1-3.

Cannon, R.C. 1960c. Pineapple research. What's in store for the grower? *Queensland Agr. J.* 86 : 635-42.

Cannon, R.C. and Prodonoff, E.T. 1959. Trials with CMU weedicides in Queensland pineapple plantations. *Queensland J. Agr. Sci.*, 16: 217-221.

Carter, W., 1933. The pineapple mealybug, *Pseudococcus brevipes*, and wilt of pineapples. *Phytopathology* 23, 207-242.

Carter, W., 1939. Populations of Thrips tabaci, with special reference to virus transmission. *Journal of Animal Ecology* 8: 261-276.

Carter, W., 1943. A promising new soil amendment and disinfectant. *Science* 97, 383-384.

Carter, W., 1967. *Insects and Related Pests of Pineapple in Hawaii.* Pineapple Research Institute of Hawaii, Honolulu, 105 pp.

Caswell, E. P. and Apt, W.J. 1989. Pineapple nematode research in Hawaii: past, present, and future. *Journal of Nematology* 21, 147-157.

Caswell, E. P., Sarah, J. L. and Apt, W.J. 1990. Nematode parasites of pineapple. In : Luc, M., Sikora, R.A. and Bridge, J. (eds) *Plant Parasitic Nematodes in Subtropical and Tropical Agriculture.* CAB International, Wallingford, UK, pp. 519-537.

Cha, J. S., Pujol, C., Ducusin, A.R., Macion, E.A., Hubbard, C.H. and Kado, C.I. 1997. Studies on *Panterea citrea*, the causal agent of pink disease of pineapple. *Journal of Phytopathology* 145, 313-319.

Chadha, K. L., K. R. Melanta, and S. D. Sikhamany. 1974. High density planting increase pineapple yield, *Indian Hart.* 18:3

Chadha, K. L., K. R. Melanta. and S. D. Sikhamany. 1973. Effect of planting density on growth, yield and fruit quality in Kew pineapple, *Indian J. Hart.* 30: 461.

Chadha, K.L. 1977. *Feasibility report on production, processing and export of pineapple products from India.* Department of Export Promotion, Ministry of Commerce, Govt of India.

Chadha, K.L., Melanta, K.R., Lodh, S.B. and Selvaraj, Y. 1972. Biochemical changes associated with growth and development of pineapple var. Kew. I. Changes in physico chemical constituents. *Indian J. Hort.* 29:54-57.

Chairidchai, P. 2000. The relationships between nitrate and molybdenum contents in pineapple grown on an inceptisol soil. *Acta Horticulturae* 529, 211-216.

Chan, Y.K, and Lee, H.K. 1996. 'Josapine': a new pineapple hybrid developed at MARDI. In: Osman, M., Clyde, M.M. and Zamrod, Z (eds) *The second National Congress on Genetics. Genetic Society of Malayasia*, UKM, Bangi.pp. 217-220.

Chan, Y.K. 1986. Hybridization and selection in F1 as a methodology for improvement of pineapple (*Ananas comosus* L. (Merr.). In: Chan, Y.K., Raveedranathan, P. and Mahmood, Z (eds) *Prosiding Simposium Buah- buahan Kabangsaan.* MARDI, Sedang, Malaysia, pp: 307-314.

Chan, Y.K. and Lee, C.K. 1985. The hybrid 1 pineapple: a new canning variety developed at MARDI. *Teknologi Buah-Buahan* 1, 24-30.

Chan, Y.K. and Lee, H.K. 2000. Breeding for early fruiting in pineapple. *Acta Horticulturae* 529,139-146.

Chang, C.C., Chang, C.C., and Chen Yung, W. 1997. Pineapple Breeding. In: Chang LinRen (ed.) *Proceeding of a Symposium on Enhancing Competitiveness of fruit industry.* Special publication no. 38. Tainchung District Agricultural Improvement Station, Taichung, Taiwan. p. 107-112.

Chang, V.C.S., and Jensen, L. 1974. Transmission of the pineapple disease organism of sugarcane by nitidulid beetles in Hawaii. *Journal of Economic Entomology* 67, 190-192.

Chen, C.C. 1999. Effects of fruit temperature, calcium, crown and sugar metabolizing enzymes on the occurrence of pineapple fruit translucency. PhD dissertation. University of Hawaii at Manoa, Honolulu, Hawaii.

Chenehin. K., A. Yugawa, and H. Y. Yamamoto. 1984. Enzymic degumming of pineapple and pineapple mill juices. *J. Food Sci.* 49: 132.

Chinchilla, C.M., Gonzales, L.C. and Morales, F. 1979. Bacterial heart rot (*Erwinia crysanthem*) of pineapple in Costa Rica. *Agronomia Cost* 3, 183-185.

Cho, J.J., Hayward, A.C. and Rohrbach, K.G. 1980. Nutritional requirements and biochemical activities of pineapple pink disease bacterial strains from Hawaii. *Antonie Van Leeuwenhoek* 46, 191-204.

Chongpraditnun, P., Luksanawimol, P. and Limsmuth-chaiporn, P. 1996. Effect of chemical, organic fertilizers and trace elements application on nitrate content in pineapple fruit. In: Vijavsegaran, S.,Pauziah, M., Mohamed, M.S. and Ahamad Tarmizi, S. (eds) *Proceedings of International Conference on Tropical Fruits*, Vol. I. Malaysian Agricultural Research and Development Institute, Kuala umpur, pp. 229-241.

Chongpraditnun, P., Luksanawimol, P., Limsmuth-chaiporn, P. and Wasunun, S. 2000. Effect of fertilizers on the content of nitrate in pineapple fruit. *Acta Horticulture* 529: 217-220.

Chowdhury, S. 1947. Pineapple culture in Assam. *Indian Farming* 8: 187-190.

Christensen, D. 1994. Performance of high density May plant crop for different graded planting material. Test No.722. In: *pineapple field day book*. Pineapple Industry Farm Committee, Beerwah, Queensland, pp. 42-48.

Christensen, D. 1995. Quality, efficiency and yield. In: *Pineapple Field Day Book*. Pineapple Industry Farm Committee, Beerwah, Queensland, pp. 15-19.

Chu, C.C. 1978. The N6 medium and its application to another culture of cereal crops [in Chinese]. In: *Proceedings of Symposium on Plant Tissue culture.* Science Press, Beijing, pp. 43-50.

Chunha, G.P.A., D.A. Coelho, Y. DA. S., and Caldas, R.C. 1980. Comnnicado Tecnico, Empresa Brasiliers de presquisa Agropecuaria, Cruzdas Almas, Behia. No. 2, 3.

Cibes, H.R. and Samuels, G. 1958. *Mineral-deficiency Symptoms Displayed by Red Spanish Pineapple Plants Grown under Controlled Conditions.* Agricultural Experiment Station Technical Paper 25, University of Puerto Rico, Rio Piedras, 32 pp.

Cibes, H.R. and Samuels, G. 1961. *Mineral-deficiency Symptom Displayed by Smooth Cayenne Pineapple Plants Grown under Controlled Conditions.* Technical Paper 31, Puerto Rico Agricultural Experiment Station, Rio Piedras, 30 pp.

Cisneros, A., Benega, R., Martinez, J., Arias, E., and Isidron, M. 1998. Effect of different plant growth regulators on the sub culture of pineapple embryogenic calli. In: *Abstracts of Third International Pineapple Symposium,* ISHS, Pattaya, Thailand, p.69.

Clark, H.E. and Kerns, K.R. 1942. Control of flowering with phytohormones.*Science* 95: 536-56.

Cobley, L.S. 1956. An introduction to the Botany of tropical crops. Longmans.

Cochereau, P. 1972a. Population dynamics of the fruit-sucking moth Otheris fullonia (Clerck) (Lepidoptera: Noctuidae) in New Caledonia. 14th *International congress of Entomology Camberra,* (Abstact) pp. 201-211.

Cochereau, P. 1977. Biologie et ecologie des populations en Nouvelle-Caledonie d'un papillon piquer de fruits: *Othreis fullonia* (Clerck). *Cahiers ORSTOM Serie Biologie* 71, 322.

Collins, J. L. 1960. *The Pineapple: Botany, Cultivation and Utilization.* Interscience Publishers, New York.

Collins, J. L., 1976. *The Pineapple: Botany, Cultivation alld Utilization.* Leonard Hill, London.

Collins, J.L. and Kerns, K.R. 1931. Genetic studies of the pineapple. I A Preliminary report upon the chromosome number and meiosis in seven pineapple varieties (*Ananas sativas* Lindl.) and in *Bromelia pinguin* L. *Journal of Heredi*ty 22, 139-142.

Collins. J. L. 1968. *The Pineapple.* World Crops Series. Leonard Hill, London.

Conway, M.J. 1977. The effects of age, temperature and duration of exposure to temperature on susceptibility of pineapple to floral induction with ethephon. MS thesis, University of Hawaii at Manoa, Honolulu, Hawaii, 102pp.

Cook, A. A., 1975. *Diseases of Tropical and Subtropical Fruits and Nuts.* Hafner Press (Macmillan), New York,

Cook, F.C. 1949. The Pineapple Indistry of the Hawaiian Islands. *Malaya Dept. of Agric. General Series* 32: 1-118.

Cooke, R. and Randall, D.I.1968. 2-haloethane phosphoric acids as ethylene releasing agents for the induction of flowering in pineapples. *Nature* 218: 974-75.

Cooper, W.C. and Reese, P.C. 1941. Inducing flowering in pineapples under Florida conditions. *Proc. Fl.St. Hort.Soc.* 54:132-38.

Coppens d'Eechkenbrugge, G., Duval, M.F. and Van Miegroet, F. 1993. Fertility and self-incompatibility in the genus *Annans. Acta Horticulture* 334- 45-51.

Coppens d'Eeckenbrugge, G. and Marie, F. 2000. Pineapple Breeding at CIRAD. II. Evaluation of Scarlet, a new hybrid for the fresh fruit market, as compared to "Smooth Cayenne'. *Acta Horticulturae* 529, 155-163.

Correa, M.P. 1952. *Diccionario das plants uties do Brasil e das exoticas cultivads,* Vol. III Imprensa Nacional, Rio de Janeiro.

Cote F., Domergue, R., Folliot, M., Bouffin, J. and Marie, F. 1961. Micropropagation in vitro de l' ananas. *Fruits* 46, 359-366.

Cowie, G.A. 1951. *Potash-Its Production* and Place *in Crop Nutrition.* London: Edward Arnold & Co.

Cox, K.J. 1979. *Decomposition Rates of Ethrel at Elevated pHs and Temperatures.* Ref note R63/ OCT 79, Queensland Department of Primary Industries, Brisbane, 2 pp.

Crafts, A.S. and Emanuelli, A. 1948. Some experiment with herbicides in pineapple. *Bot. Gaz.* 110: 312-319.

Crochon, M., Tisseau, R., Teisson, C. and Hue, R.1981. Effect of preharvest application of ethrel on the flavour of pineapples in the Ivory Coast. *Fruits* 36:40-15.

Dalldorf, D.B. 1979. *Flower Induction of Pineapples*. Department of Agricultural Technical Services, Farming in South Africa, Pretoria, 8 pp.

Dalldorf, D.B. and Langenegger, W. 1976. *The Influence of Potassium on the Yield, Fruit Quality, and Plant Growth of Smooth Cayenne Pineapples* Gewasproduksie, Pretoria, 5 pp.

Daquinta, M. Cisneros, A., Rodriquez, Y., Escalona, M., Perez, C., Luna, I. and Borroto, C.G. 1996. Somatic embryogenesis in pineapple (*Ananas comosus* (L.) Merr.). In: Pineapple Working Group Newsletter. *International Society for Horticulture Science*, Honolulu, Hawaii, pp-5—6.

Das, H. 1964. Studies on the action of NAA on the flowering and fruiting of pineapple. *Indian J. Agric. Sci.* 34: 38-45.

Das, N., Baruah, S.N. and Baruah, A. 1965. Induction of flowering and fruit formation of pineapples with the aid of acetylene and carbide. *Indian Agriculturist* 9: 15-23.

Dass, H.C., Ganapathy, K.M., Singh, H.P., Reddy, B.M.C and Prakash, G.S. 1978. Spacing and population density studies in pineapple var.Kew. Paper presented at *All India Coordinated Fruit Improvement Project*, held at Bangalore.From Research Report and Project Proposals on Banana, Pineapple, Papaya, pp. 130-32.

Dass, H.C., Randhawa, G.S, G.S. and Negi, S.P. 1975. Flowering in pineapple as influenced by ethrel and its combinations with urea and calcium carbonate. *Scientia Horticulturae* 3:231-38.

Dass, H.C., Singh, H.P., Ganapathy, K.M and Randhawa, G.S. 1977. Standardisation of optimum leaf number for induction of flowering pineapple. *Indian J. Hort* 34:24-25

Dass, H.C., Sohi, H.C., Reddy, B.M.C and Prakash, G.S. 1976. Vegetative multiplication by leaf cuttings of crowns in pineapple [*Ananas comosus* (L). Merrill]. *Current Science* 46: 241-2.

De Geus, J. G. 1973. Fertilizer Guide for the Tropics and Sub tropics. Zurich.

De Geus. J.G. 1961. Fertilizer requirements for sugarcane, deciduous fruits, citrus and pineapple in the Union of south Africa. Stikst of (Eng. Edition) 5: 53-68.

Devi, Y.S., Mujib, A and Kundu, S.C. 1997. Efficient regenerative potential from long term culture of pineapple. *Phytomorphology* 47, 255-259.

DeWald, M.G., Moore, G.A. and Sherman, W.B. 1992. Isozymes in Ananas (pineapple): genetics and usefulness in taxonomy. *Journal of American Society of Horticultural Science* 117(3), 491-496.

Dhareswar, S.R. 1950. Trench cultivation of pineapple. *Dharwar Agric. Col. Mag.* 4: 9-11.

Drew, R.A. 1980. Pineapple tissue culture unequalled for rapid multiplication *Queensland Agriculture Journal* 106, 447-451.

Dull, G.G. 1975. The Pineapple, *The Biochemistry of Fruits and their Products*, Vol. 2 (A. C. Hulme, ed.), Academic Press, London and New York, pp. 303-324.

Dunsmore, J.R. 1957. The pineapple in Malaya (*Anana comosus* L. Merr.) *Malayan Agr. J.* 40: 159-87.

Dunsmore, J.R. 1957. The pineapple in Malaya (*Ananas comosus* (L.) Merr.) Malayan Agr. J. 40: 159-87.

Ekern, P.C. 1965. *Plant Physiol.* 40, p. 736.

Escalona, M., Lorenzo, J.C., Gonzalez, B., Daquinta, M., Fundora, Z., Borroto, C.G., Espinosa, D., Arias, E. and Aspiolea, E. 1998. New System for in-vitro propagation of pineapple [*Ananas comosus* (L.) Merr.]. *Tropical Fruits Newsletter* 29,3—5.

Evans, H.R. 1959. The influence of growth promoting substances on pineapples. *Tropical Agriculture, Trinidad* 36: 108-117.

Ewing. C.B., R.P. Muns and H.J. Kaplan. 1980. Lengthening storage life of pineapples through the use of selected coating materials (Abstr.). *Hort Sci.* 15(3): 93.

F.A.O. *Production Year Book*. Vol. 34, Food and Agriculture Organization. Rome, 1990.

Firoozabady, E., Nicholas, J. and Gutterson, N. 1995. *In vitro* plant generation and advanced propagation methods for pineapple. In: *Abstracts of the Smooth Cayenne Pineapple Symposium*. ISHS, Fortde-France, Martinique, p.12.

Fitchet, M. 1989. Observations on pineapple improvement in Taiwan, Republic of China. *Subtropica* 10-, 10-12.

Fitchet, M. 1990. Clonal Propagation of Queen and Smooth Cayenne pineapples. *Acta Horticulture* 275, 261-266.

Fitchet-Purnell, M. 1993. Maximum utilization of pineapple crowns for micropropagation. *Acta Horticulture* 334,325-330.

Fleish, H. and Bartholomew, D.P. 1987. Development of a heat unit model of pineapple Smooth Cayenne) fruit growth from field data. *Fruits* 42: 709-715.

Follett-Smith, R.R. and Bourne, C.L. 1936. The uptake of minerals by pineapple plants at different stages of growth. *Agric. J. Brit.* Guyana 7:17-20.

Friend, D.J.C. and Lydon, J. 1979. Effects of day length on flowering, growth, and CAM (crassulacean acid metabolism) of pineapple [*Ananas comosus* (L). Merrill]. *Botanical Gazette* 140,280-283.

Funasaki, G.Y., Lai, P.Y., Nakahara, L.M., Beardsley, J.W. and Ota, A.K. 1988. A review of biological introductions in Hawaii: 1980-1985. *Proceedings of the Hawaiian Entomological Society* 28, 105-160.

Gaillard, E.J., Py, C. and Lossois, P. 1984. Estudo sobre o ciclo natural do abacaxizeiro 'Cayenne' no planalto paulista. *Bragantia Campinas* 43,629-642.

Gaillard, J.P. (1969) Influence of planting date and weight of slips on the growth of pineapples in Cameron. *Fruits* 24, 75-87.

Gaillard, J.P.1971. Control of *Cyperus rotundus* in pineapple. *Fruits* 26 : 751-56.

Gandhi, S.R. 1949. Pineapple culture in Western India.*Poona Agric. Col. Mag.* 40(3): 33-39.

Ganpathy, K.M., Singh, H.P. and Das, H.C. 1977. A note on phynotypic expression of some characters in Kew pineapple influenced by nutritional levels of the plants. *Indian J. of Agric. Sci.* 34: 142-43.

Garcia, M. and Treto, E. 1988. Phosphorus fertilization and its residual effect on pineapple cultivar Spanish Red. *Cultivos Tropicales* 10: 74-79.

George, P.V. and Ooman, M. 1968. Some striking freaks in *Ananas comosus* L. *Indian J. Hort.* 25 : 178-79.

Giacomelli, E.J. 1967. The pineapple plant and water. *Agronomico Companas* 19: 4-68.

Gilmartin A.J. and Brown, G.K. 1987. Bromeliales, related monocots, and resolution of relationships among Bromeliaceae subfamilies. *Systematic Botany*. 12, 493-500.

Glennie, J.D. 1977. *Pineapple Nutrition*. Advisory Leaflet H34, Horticultural Branch, Queensland. Department of Primary Industries, Namhour, Queensland, Australia.

Glennie, J.D. 1979. The effect of temperature on flower induction of pineapple with Ethrel. In: *Pineapple Field Day Book*. Pineapple Industry farm Committee, Beerwah, Queensland, pp.8-13.

Godefroy, J. 1979. *Erosion et pertes par lixiviation on nuissellement des elements fertilisants sous culture d'ananas enfonction des technique* culturales*. Document interne no. 6, IRFA, Monrpellier, France.

Godefroy, J. Poignant, A. and Marchal, J. 1971. First results from a pineapple liming trial on a lower Ivory Coast soil. *Fruits 26*, 103-113.

Gonzales R., Dominguez, Q., Exposition, L., Jorge, L., Martinez, T., and hidalgo, M. 1996. Effectiveness of 8 strains of Azobacter on the adaption of tissue cultured plantlets of pineapple (*Ananas comosus* Merr.) cv. 'Smooth Cayenne'. *Acta Horticulture* 425,277-284.

Gonzalez, Tejera, E. and Gandia Diaz, H. 1976. The effects of nitrogen and potassium fertilizers on the productivity and quality of pineapple cultivar Smooth Cayenne. In" *24th Annual Congress of the Amer. Society Hort. Sci., Tropical Region*, pp. 196-205.

Gonzalez-Hernandez, H., Johnson, M.W. and Reimer, N.J. 1999. Impact of *Pheidole megacephala* (F) (Hymenoptera: Formicidae) on the biological control of *Dysmicoccus brevipes* (Cockerell) (Homoptera: Pseudococcidae). *Biological Control: Theory and Applications in Pest Management* 15, 145-152.

Gortner, W.A. and Singhleton, V.L., (1965. Chemical and physical development of the pineapple fruit. III. Nitrogenous and enzyme constituents. *Journal of Food Science* 30, 24-29.

Gortner, W.A., (1969. Relation of chemical structure to plant growth-regulator activity in the pineapple plant: reloading senescence of pineapple fruit with applications of 2, 4, 5- trichlorophenoxy acetic acid and 1-naphthalene acetic acid. *Journal of Food Science* 34, 577-580.

Gortner, W.A., Dull, G.G. and Krauss, B.H. 1967. Fruit development, maturation, ripening and senescence: a biochemical basis for horticultural terminology. *Hortscience*, 2: 141-144.

Gossele, F. and Swings, J. 1986. Identification of *Acetobacter liquefaciens* as casual agent of pink-disease of pineapple fruit. *Journal of Phytopathology*, 116: 167-175.

Gowing, D.P. 1961. Experiments on the photoperiodic response in pineapple. *American Journal of Botany*, 48: 16-21.

Gowing, D.P. and Leepar, R.W. 1955. Induction of flowering in pineapple by betahydroxy- ethyl hydrazine. Science 122:1267.

Gowing, D.P. and Leeper, R.W. 1959. *Further Data comparing BOH and SNA as Forcing Agents. PRI* Research Report No.63, private document, Pineapple Research Institute of Hawaii, Honolulu.

Gowing, D.P. and Leeper, R.W. 1960. Studies on the relation of chemical structure to plant growth- regulator activity in the pineapple plant I. Substituted phenyl and phenoxyalkylcarboxylic acids. *Botanical Gazette* 121,143-151.

Griffin, W. 1806. A Treatise on the Culture of the Pineapple. Printed for the author by Ridge, S. and Ridge, J., Newark, UK.

Guerout, R. 1975. Nematodes of pineapple a review. *Pans* (*Pest Article News Summary*) 21, 123-140.

Guillemin, J.P., Gianinazzi, S. and Gianinazzi-Pearson, M. 1996. Endomycorrhiza biotechnology and micropropagated pineapple (*Ananas comosus* (L.) Merr.). *Acta Horticultute* 425, 267-275.

Guillemin, J.P., Gianinazzi, S. and Gianinazzi-pearson, V. 1997. Endomycorrhiza biotechnology and micropropagated pineapple (*Ananas comosus* (L.) Merr). *Acta Horticulture* 425-267-275.

Guyot, A. and Py, C. 1970. Controlled the flowering of pineapples with Ethrel, a new growth regulator (Conclusion). *Fruits* 25,427-445.

Guyot, H. and Oliver, P. 1958. Herbicides per atomization (Low volume application of herbicides). *Fruits d'Outer Mer.*13 : 203-208.

Hariprakasa Rao, M., Bubramanian, T.R., Srinivasa Murthy, H.K. Singh, H.P. Das, H.C and Ganapathy, K.M. 1977. Leaf nitrogen status as influenced by varying levels of nitrogen application and its relationship with yield in 'Kew" pineapple. *Scientia Horticulturae,* p. 137-42.

Hartung, W.J., Magistad, O.C. and Thot, K. 1931. Calcium treatment of acid soild. *Pineapple Quarterly* 1, 139-145.

Hayes, W. B. 1959/60. Fruit Growing in India. Kitabistan, Allahabad, pp.368-81.

Heenkendra, H.M.S. 1993. Effect of plant size on sucker promotion in 'Mauratius' pineapple by mechanical decapitation. *Acta Horticulturae* 334: 331-336.

Henke, L.A. 1934. *Pineapple Plants as Forage for Cuttle.* Progress Notes No. 6, Hawaii Agricultural Experiment Station, Honolulu, 81 pp

Hine, R. B., Alaban, C. and Klemmer, H.W. 1964. Influence of soil temperature on root and heart rot of pineapple caused by *Phytophthora cinnamomi and Phytophthora parasitica. Phytopathology* 54, 1287-1289.

Hine, R.B., (1976. Epidemiology of pink disease if pineapple fruit. *Phytopathology* 66, 323-327.

Hinton, H.E. 1945. *Monograph of Beetles Associated with Stored Products.* Jarold and Sons, London.

Hopkins, E.F, Pagan, V. and Silva, F.J.R. 1944. Iron and manganese in relation to plant growth and its importance in Puerto Rico. *Journal of Agriculture, University of Puerto Rico* 28, 43-101.

Hu, J.S., Gonsalves, A., Sether, D. and Ulman, D.E. 1993. Detection of pineapple clostero virus, a possible cause of mealybug wilt of pineapple. In: Bartholomew, D.P. and Rohrbach, K.G. (eds) *First International Pineapple Symposium. Acta Horticulturae,* Honolulu, Hawaii, pp. 411-416.

Huang, C.C. and Lee, L.C. 1969. Effect on irrigation of pineapple. *Taiwan Agric. Quart.* 5: 50-8.

Huet, R. and Tisseau, M.A. 1959. Observations on the development of pineapple after harvesting. *Fruits d'Outre Mer.* 14: 271-74.

Hume, H.H. and Miller, H.K. 1904. Pineapple culture II: varieties. *Bull. Fla. Agric. Exp Sta.* 70: 37-62.

Iglesias, R. 1979. Influence of a mixture of ethrel with urea and sodium carbonate on flowering, fruit quality and slip production in the pineapple cultivar. Espanola Roja. *Cultivos Tropicales* 1,117-130.

IIFT. 1968. Survey of India's export potential of fresh and processes fruits and vegetables. Indian *Institute of Foreign Trade* 6(IB): 385, 404-12.

IIHR, Bangalore, 1977. Pineapple cultivation.Extension bulletin No.6. Indian Institute of Horticultural Research, Bangalore.

Illingworth, J.F. 1931. Yellow spot of pineapples in Hawaii. *Phytopathology 21, 865-880.*

Ingamells, J.L. 1981. The effects of pineapple residue (trash) on mineralization and early growth of pineapple. PhD dissertation, University of Hawaii at Manoa, Honolulu, Hawaii

Jappson, L.R., Keifer, H.H. and Baker, E.W. 1975. *Mites Injurious to Economic Plants.* University of California Press, Berkeley, California.

Johnson, A.W. and Feldmesser, J. 1987. Nematicides: a historical review. In: Veech, J.A.and Dickson, D.W. (eds) *Vistas on Nematology: A Commemoration of the Twenty-fifth Anniversary of the Society of Nematologists.* Society of Nematologists, Hyattsville, Maryland, pp. 448-454.

Joseph. G. and Mahadeviah, M. 1988. Utilization of waste from pineapple processing industries. *Indian Food Packer* 42: 46.

Joshi, M.C. *et al.*, 1965. *Bot. Gaz.* 126. p. 174.

Kanapathy, K. 1959. Visual symptoms of major nutrient deficiencies of the Singapore Spanish pineapple. *Malnysian Agriculture Journal*, 42: 157-160.

Katsumi, U., H. Yukio, N. Kazwaki, S. Akihiro, and S. Takayuki. 1992. Volatile constituents of green and ripened pineapple. *J. Agril. Food Chem.*, 40: 599.

Keetch, D.P. 1977. H.1 *Black Spot (Fruitlet Core Rot) in Pineapple*. Pine Series H: Diseases and Pests, Government Printer, Pretoria, Republic of South Africa, 3pp.

Keetch, D.P. 1979. H. 19 *Pineapple–Nematode Control with Fumigant Nematicides*. Department of Agricultural Technical Services, Pretoria, South Africa, 4pp.

Kerns, K.R. 1935. Method and material for forcing flowering and fruit formation in plants. *U.S. Plant Patent* 2,047,874.

Kerns, K.R., Collins, C.S. J.L. and Kim, H. 1936. Development studies of the pineapple *Ananas comosus* (L) Merr. *The New Phytologist* 35: 305-317.

Kerns, K.R., Collins, J.L. and Kim, H. 1936. Developmental studies of the pineapple *Ananas comosus* (L.) Merr.I.origin and growth of leaves and inflorescence. *New Phytologist* 35: 305-317.

Ketch, D.P., (1979. H.19 *Pineapple—Nematode Control with Fumigant Nematicides*. Department of Agricultural Technical Services, Pretoria, South Africa, 3 pp.

Khatua, N., Mitra, S.K. and Bose, T.K. 1988. Influence of various nitrogenous fertilizers on yield and quality of Kew pineapple. *Haryana J. Hort. Sciences* 17: 190-93.

Khongwir, R.C. and Das, D. 1978. Manurial-cum-spacing trial on pineapple var.Giant Kew. Paper presented at *Fruit Research Workshop held at the University of Agricultural Sciences*, Bangalore.

Kiss, E., Kiss, J., Gyulai, G. and Heszky, L.E. 1995. A novel method for rapid micropropagation of pineapple. *HortScience* 30,127-129.

Klemmer, H.W. and Nakano, R.Y. 1964. Distribution and pathogenicity of *Phytophthora and Pythium in* pineapple soils in Hawaii. *Plant Disease Reporter* 48: 848-852.

Kontaxis, D. G. 1978. Control of pink disease of pineapple fruit with disulfoton in the Philippines. *Plant Disease Reporter* 62, 172-173.

Kontaxis, D.G. and Hayward, A.C. 1978. The pathogen and symptomatology of pink disease, *Acetobacter aceti, Gluconobacter oxydans, on pineapple fruit in the Philippines. Plant Disease Reporter* 62, 446-450.

Kraus, B.H. 1949b. Anatomy of the vegetative organs of the pineapple, *Ananas comosus* (L.) Merr. Concluded. III. The root and the cork. *Botanical Gazette* 110, 550-587.

Krauss, F.G. 1928. Die Chemische Zusammensetzung der Ananaspflanze in verschiedenen Wachstumsstadien. *Ernahr. d. Pflanze* 24: 398.

Kwong, K.H. and Chiu, Y.M. 1968. Use of SNA sprays to improve weight, size, spring fruit of pineapples in the Taitung district. *J. Agric. Ass. China* 9:39-43.

Lacoeuihe, J.J. 1978. La fumure N-K de l'ananas en Cote d' Ivoire. *Fruits* 33,341-348.

Lacoeuilhe, J.J. 1973. Rythme d' absorption du potassium en relation avec in croissance : cas de l' ananas et du bananier. 10. Colleque de l' Institut International de la pottase, Abidjan, 177-83.

Lacoeuilhe, J.J. 1976. Croissance de l' ananas en function du type de reject et de la fumure. Bilans en matiere fraiche et seche et en elements mineraux. Reunion Annuelle. IRFA, doc. Interne, no.11.

Lal, G. and Pruthi, J.S. 1955. Ascorbic acid retention in pineapple products. *Indian J. Hort.* 12:137-41.

Langenegger, W. and Smith, B.L. 1978. An evaluation of the DRIS system as applied to pineapple leaf analysis. "Plant Nutrition" *Proceedings of the 8th Int. Colloquium on Plant Analysis and Fertilizer Problems.* Auckland, News Land, 28th August -1 September. 1978.

Laufer, B. 1929. The American Plant Migration. *Science Monthl.y* 28, 239-251.

Laville, E. 1980. *Fusarium* disease of pineapple in Brazil. I. Review of current knowledge. *Fruits* 35, 101-113.

Le Grice, D.S. and Marr, G. S. 1970. Fruit disease control in pineapple. *Farming in South Africa* 46, 9, 12, 17.

Leal, F and Coppens d' Eeckenbrugge, G. 1996. *Pineapple.* In Janick, J. and Moore, J. N. (eds) *Fruit Breeding.* John Wiley and Sons. New York, pp. 565-606.

Leal, F. and Amaya, L. 1991. The Curagua (*Ananas lucidus*, Bromeliaceae) crop in Vanizuela. *Economic Botany.* 45(2). 216-224.

Leal, F. and Coppens d'Eeckenbrugge, G. 1996. Pineapple In: Janick, J. and Moore, J.N.(eds) *Fruit Breeding.I. Tree and Tropical Fruits.Vol.I.* John Wiley and Sons, New York, pp. 515-557.

Leal, F., Coppens d'Eeckenbrugge, G and Holst, B.K. 1998. Taxonomy of Genera Anas and Pseudananas: A historic review. *Selbyana* 19, 227-235.

Leeper, R.L. 1965. Factors influencing forcing and delaying: a review. *Pineapple News* 13,109-121. Private document, Pineapple Research Institute of Hawaii, Honolulu.

Leme, E.M.C. and Marigo, L.C. 1993. Bromiliads in the Brazilian Wilderness. Marigo Comunicacao Visual, Rio de Janerio.

Lewcock, H.K. 1937. The use of acetylene to induce flowering in pineapple plants. *Queensland Agriculture Journal* 48,532-543.

Lim, W. H. and Lowings, P. H. 1979a. Effects of ethephon on anthesis and 'fruit collapse' disease in pineapple. *Experimental Agriculture* 15, 331-334.

Lim, W.H. 1971. Evaluating the susceptibility of pineapple cultivars to bacterial heart rot. *Malaysian Pineapple* 1, 23-27.

Lim, W.H. 1985. *Diseases and Disorders of Pineapples in Peninsular Malaysia.* MARDI Report No. 97, Malaysian Agricultural Research and Development Institute (MARDI), Kuala Lumpur, Malaysia, 53 pp.

Lim, W.H. and Lowings, P.H. 1978. Infections sites of pineapple fruit collapse and latency of the pathogen, *Erwinia chrysanthemi,* within the fruit. In: *Proceedings 4th International Conference Plant Pathogenic Bacteria, Angers,* France, pp. 567, 575.

Lim, W.H. and Lowings, P.H. 1979b. Pineapple fruit collapse in Peninsular Malaysia: symptoms and varietal susceptibility. *Plant Disease Reporter* 63, 170-174.

Lindeley, J.J. 1827. *Billbergia.* Botanical Register 13, 1068.

Linden, M.J. 1879. *Ananas Mordilona* Linden. Belgique *Horticole* 29, 302-303.

Linford, M.B. 1932. Transmission of the pineapple yellow-spot virus by *Thrips tabaci. Phytopathology* 22, 301-324.

Linford, M.B. 1943. Influence of plant populations upon incidence of pineapple yellow spot. *Phytopathology* 33, 408-410.

Linford, M.B. and Spiegelberg, C.H. 1933. Illustrated list of pineapple fruit disease, blemishes and malformalities. *Pineapple Quaterly*. 3(4): 134-178.

Linford, M.B., Oliveira, J. M. And Ishii, M. 1949. *Paratylenchus minutus, n. sp.*, a nematode parasitic on roots. *Pacific Science* 3, 111-119.

Linford,M.B. 1933. Fruit quality studies. II. Eye number and eye weight. *Pineapple Quarterly* 3,185-188.

Linnaeus, C. 1753. *Species plantarum*. Stockholm Sweden, 724 pp.

Lodh, S.B, Diwakker, N.G., Chadha, K.L. Melanta, K.R. and Selvaraj, Y. 1973. Biochemical changes associated with growth and development of pineapple fruits var. Kew. III. *Indian J Hort*. 30: 381-383.

Lodh, S.B., Selvaraj, Y., Chadha, K.L., Melanta K.R. 1972. Biochemical changes associated with growth and development of pineapple fruits var. Kew. II. Change in carbohydrates and mineral constituents. *Indian J. Hort*. 29:287-92.

Loison-Cabot, C. 1987. Practice of pineapple breeding. *Acta Horticultarae*. 196, 25-36.

Luther, H.E. and Sieff, E. 1998. An *Alphabetical List of Bromeliad Bionmials*, 6ᵗʰ edn. The Biomeliad Society, Newberg, Oregon, 138 pp. http://www. Selby.org./research/bic/lino98.htm.

Lyon, H.L. 1915. A Survey of pineapple problems. *Planters Record* 13, 125-139.

Maffia, L. A. 1980. Persistence of *Fusarium moniliforme var. subglutinans* in the soil and on crop residues and its elimination from pineapple suckers by hot water treatment. *Fruits* 35, 217-243.

Magistad, O.C., Fardent, G.A. and Baldwin, W.A. 1935. Bagasse and paper mulches. *J. Amer. Soc. Agron*.27: 813-25.

Maity, S.C. and Sambui, D.N. 1980. Pineapple production and utilization, *Proc. Nat. Sem*. Calcutta, pp. 52-57

Malan, E.F. 1954. Pineapple production in South Africa. Fmg S Afr. 29 : 175-180.

Malezieux, E. 1988. Croissance et elaboration du rendement de I'ananas (*Ananas comosus* L. Merr). *Doctorat en Sciences*, INA–PG, Paris.

Malezieux, E. and Sebillotte, M. 1990a. *Relations entre les processus d'accumulation de la matiere seche et le rendement chez l'ananas* (Ananas comosus L. *Merr.*). II. *L'effet de la biomasse presente induction florale.* Note dereunion Annuelle IRFA, document interne, Moutpellier.

Malezieux, E. and Sebillotte, M. 1990b. *Relations entre les processus d'accumulation de la mateire seche et le rendement chez l'ananas (Ananas comosus* L. Merr.). IV. *L'effet du climat.* Note de reunion Annuelle IRFA, document interne, Montpellier.

Malezieux, E., Zhang, Sinclair, E. and Bartholomew, D.P. 1994. Predicting pineapple harvest date in different environments, using a computer simulation model. *Agronomy Journal* 86,609-617.

Mangaraj, B.K. 1981. Studies on utilization of pineapple juice for concentration. M.Sc. Food Tech. Investigation Rep. Central Food Technology Research Institute, Mysore, India.

Manuel, F.C. 1962. Control of weeds in pineapple with two soil-applied herbicides. *Philip. Agr.* 46: 514-24.

Mapes, M.O. 1973. Tissue culture in bromeliads. *The International Plant Propagators Society* 23: 47-55.

Marchal, J. and Pinon, A. 1980. Nutrition azotee de Pananas. Etude des voies d' absorption de l' azote for la dilution isotopique. *Fruits* 35:29-38.

Martin Prevel, P. 1959. Apercu sur les relations croissance-nutrition minerale chez l' ananas. *Fruits* 14, 101-122.

Martin Prevel, P. 1970. Practical uses of foliar diagnosis. *Fruits* 25, 117-123.

Martin-Prevel, P. 1959b. Carence en potassium sur ananas en Guinee (Potassium deficiency in pineapples in Guinea). Fruits d'outre Mer. 14: 285-89, 414-19.

Martin-Prevel, P. 1961a. Potassium, calcium et magnesium dans la nutrition de lnanas en Guinee. I.Plan et deroulement de 1'etude. *Fruits* 16, 49-56.

Martin-Prevel, P. 1961b. Potassium, calcium et magnesium dans la nutrition de l'ananas en Guinee. II. Influence sur le rendement commercialisable. *Fruits* 16, 113-123.

Martin-Prevel, P. 1961c. Potassium, calcium et magnesium dans la nutrition de l'ananas en Guinee. IV.Etude de la croissance foliaire. *Fruits* 16, 341-351.

Martin-Prevel, P. 1961d. Potassium, calcium et magnesium dans la nutrition de l'ananas en Guinee. V. Donnees de 1'analyse foliaire. *Fruits,* 16, 539-557.

Martin–Prevel, P. *et al.,* 1959a. Echantillonage de l' ananas en diagnostic foliare. Nutrition minerale en engrais. Abidjan 16-22/2/59, p. 57.

Martin–Prevel, P. *et al.,* 1962. Potasium, calcium et et magnesium dans la nutrition de l' amamas Guinee. *Fruits* 17: 211-27, 257-61.

Martin-Prevel, P., Huet, R. and Haendler, L. 1961. Potassium, calcium et magnesium dans la nutrition de l'ananas en Guinee. III. Influence sur la qualite du fruit. *Fruits* 16, 161-180.

Mathews, V.H., and Rangan, T.S. 1979. Multiple plantlets in lateral bud and leaf explants in vitro cultures of pineapple. *Scientia Horticulture* 11,319-328.

Mathews, V.H., Rangan, T.S. and Narayanaswamy, S. 1976. Micropropagation of *Ananas sativus in vitro. Pfanzenphsiol.* 79: 445-55.

Matos, R.M.B., Da-Silva, E.M.P. and Da-Silva, E.M.R. 1996. Effect of inoculation by arbuscular mycrorhizal fungi on the growth of micropropagated pineapple plants. *Fruits* 51(2), 115-119.

Melo, S.A.P.d., Araujo, E. and Suassuna, J. 1974. Preliminary note on the occurence of *Eraraizia* sp. Causing a dry rot of pineapple fruits in Paraiba. *Review of Agriculture* 49, 124.

Mez, C. 1892. Bromeliacease, *Ananas: Martius, Flora Brasiliensis,* Vol 3 (3). Reprinted 1965 Verlag von J. Cramer, Weinheim, Codicote (hertfordshire), Wheldon and Wesley, New York, pp. 288-294)

Miles Thomas, E.N. and Holmes, L.E. 1930. The development and structure of the seedling and Young plants of the pineapple (*Ananas sativus*). New Phytologist 29, 199 226.

Millar-Watt, D. 1981. Control of natural flowering in the Smooth Cayenne pineapple, *Ananas comosus* (L.) *Merr. Citrus and subtropical Fruit Research Institute Bulletin* 110, 17-19.

Miller, P. 1754. *Gardener's Dictionary, 4*[th] *edn.* Henrey, Staflen and Cowan, London.

Miller, P. 1768. *Gardener's Dictionary, 8*[th] *edn.* Henrey, Staflen and Cowan, London.

Min, X-J. 1995. Physiological effects of environmental factors and growth regulators on floral initiation and development of pineapple, *Ananas comosus* (L.) Merr. PHD dissertation, University of Hawaii at Manoa, Honolulu, Hawaii.

Min, X-J. and Bartholomew, D.P. 1996. Effect of plant growth regulators on ethylene productions, 1-aminocyclopropane-1-carboxylic acid oxidase activity, and initiation of inflorescence development of pineapple. *Journal of Plant Growth Regulation* 15: 121–128.

Min, X-J. and Bartholomew, D.P. 1997. Temperature affects ethylene metabolism and fruit initiation and size of pineapple. *Acta Horticulturae* 425,329-338.

Mitchell, P. and Nicholson, M.E. 1965. Pineapple growth and yield as influenced by urea spray schedules and potassium levels at three plant spacings. *Queensland J. Agric. Anim, Sci.* 22:409-17.

Mol'. J. H., E. Ross. S. T. Hsia. and J. Sawato. 1967-68. Quality evaluation of gamma irradiated papaya in shipping studies. U.S. Atomic Energy Commission. Di, . of Isotope Development. *Ann. Rep.,* p. 112.

Montinola, L.R. 1991. *Pina* Amon Foundation. Manilla Philippines. Oviedo Gonzalo F. De. 1535. Historia naturally general de las Indias. Ed. Atlas Madrid. 1959, vol. 5.

Moore, G.A., DeWald, M.G. and Evans, M.H. 1992. Micropropagation of pineapple (*Ananas comosus* L.). In: Bajaj, Y.P.S. (ed.) *Biotechnology in Agriculture and Forestry,* Vol. 18, *High-Tech* and *Micropropagation II.* Springer-Verlag, Berlin, pp.460-470.

Morgan, P.W., He, C.J., De Greef, J.A. and proft, M.P. 1990. Does water deficit stress promote production potential of Colombia. *Agricultural Meteorology* 17, 81-92.

Morren, E. 1878. Description, de I, Ananas macrodontes. Sp. Nov. Ananas a fortes epines. *Belgique Horticole (Liege)* 28, 140-172.

Morrison, S.E. 1963. Journals and other documents of the life and voyages of Christopher Columbus. Heritage Press, N. York.

Mukherjee, S.k., Rao, D.P., Das, C.S and Saha, P.K. 1982. Effect of planting density on growth, yield and fruit quality of pineapple cv.Queen *Ananas comosus* (L.) Merr. *Indian J. Hort.* 39: 1-8.

Munro, D. 1985. Classification of pineapple varieties. *Transactions of London Horticultural* Society 1, 1-34.

Murashige, T. and Skoog, F. 1962. A revised medium for rapid growth and bioassays with tobacoo tissue cultures. Physiologia Plantarum 15,473-497.

Murashige, T. and Tucker, D.P.H. 1969. Growth factor requirements of citrus tissue culture. In: chapman, H.D. (ed.) *Proceedings of the First International Citrus Symposium.* University of California-Riverside Publication Riverside, California, pp. 1155-1161.

Mustaffa, M.M. 1989. Effect of phosphorus application on fruit yield, quality and leaf nutrient content of Kew pineapple. *Fruits* 44, 253-257.

Naik, K.C., 1963. *South India Fruits and their Culture,* P. Varadachary and Co., Madras, pp. 303.

Nanjundaswamy, A. M. K. C. Chik.kappaji. and M. V. Patwardhan. 1980. Studies on processing of pineapples, *Proc. Seminar on Pineapple and its Utilization.* Associated Food Scientists and Technologists of India. Oct. 4-5, Jadhavpur University, Calcutta.

Neild, R.E. and Boshell, F. 1976. An agro climatic procedure and survey of the pineapple production potential of Colombia. *Agricultural Meteorology* 17, 81-92.

Nforzato, R. *et al.* 1968. *Bragantia* 27, p. 135.

Nightingale, G.T. 1942a. Nitrate and carbohydrate reserves in relation to nitrogen nutrition of pineapple. *Botanical Gazette,* 103: 409-456.

Nitsch, J.P. 1951. Growth and development *in vitro* of excised ovaries. *American Journal of Botany* 38,566-577.

Norman, J.C. 1978. Responses of 'sugarloaf' pineapple, *Ananas comosus* (L.) Merr to plant population densities. *Gartenbauwissenschafte,* 43: 237-24.

Nunez Soto, A. and Garcia Serrano, T. 1978. Influence de distintas formas de application del fertilizante NK sabre el cultivo de la pina [*Ananas comosus*) (L.) Merr.]. *Ciencia Y. Tecmia en la Agriclture, Suelos Y. Aroquinicia*, 1(1): 19-29.

O'Donnell, K., Cigelnik, E. and Nirenberg, H.I. 1998. Molecular systematic and Physiography of the *Gibberella fujikuroj* species complex. *Mycologia* 90, 465-493.

Okimoto, M.C. 1948. Anatomy and Histology of the pineapple inflorescence and fruit. *Botanical Gazette* 110, 217-231.

Osburn, M.R. 1945. Methyl bromide for control of pineapple mealybug. Journal of Economic Entomology. 38,610.

Osei-Kofi, F. and Adachi, T. 1993. Effect of cytokinins on the proliferation of multiple shoots of pineapple in vitro. Sabrao Journal 25,59-69.

Ota, S., Muta, F., Katahira, Y. and Okamoto, Y. 1985. Reinvestigation of fraction and some properties of the proteolytically active components of stem and fruit bromelains. *Journal of Biochemistry* 98, 219-228.

Pablo, I.S. Akamine, E.K. and Chachin, K. 1975. Irradiation. *Postharvst Physiology of Subtropical Fruits and Vegetables* (E.B. Pantastico. ed.). AVI. Westpoll. CT. pp. 219-235.

Palmer, R.L., Lewis, L.N., Hield, H.Z and Kumamoto, J. 1967. Abscission induced by b-hydroxyethylhydrazine: conversion of b-hydroxyethylhydrazine to ethylene. *Nature* 216, 1216-1217.

Pan, K.Y. 1957. Research on the three essential fertilizers for pineapple. *J. Agric. Assoc. China*, p. 11-129.

Pannetier, C. and Lanaud, C. 1976. Divers aspects de l'utilization possible des cultures in vitro pour la multiplication vegetative de l' ananas. *Fruits* 31,739-750.

Paull, R.E. and Rohrbach, K.G. 1985. Symptom development of chilling injury of pineapple fruit. *Journal of the American Society for Horticultural Science*, 110: 100-105.

Paull, R.E., and Rohrbach, K.G. 1982. Juice characteristics and internal atmosphere of waxed 'Smooth Cayenne' pineapple fruit. *Journal of the American Society for Horticultural Science* 107, 448-452.

Pena Arderi, H. and Dominguez Martin, Q. 1988. Materia Organica y nutrients que se incorporan al suelo can la demolicion de lal pina, *Ananas comosus* (L.) Merr., cultivar Espanola Roja (Organc mater and nutrients added to the soil after grubbing a pineapple plantation, *Ananas comosus* (L.) Merr. Cv. Spanish red). *Centro Agricola* 15,84-87.

Perrotet, S. 1825. Catalogue raisonne des plantes introduites dans les colonies francaises de Mas careigne et de Cayuenne, et de celles rapports'ees vivantes des mersd'Asie et de la Guyane, au Jardin des Plantes de Paris. *Memoires Society Linneas.* 3(3), 89-151.

Petty, G. and Webster, G. 1979. False spider mite control. *Information Bulletin Citrus Subtropical Research Institute,* 84, 3-4.

Petty, G.J. 1975. Pineapple mites. *Citrus Subtropical Fruit Journal,* 498: 15-18.

Petty, G.J. 1977b. *H.2 Pineapple Pests: Leathery Pocket in Pineapple.* Pineapple Series II, Diseases and Pests Government Printer, Pretoria, republic of South Africa, 3pp.

Petty, G.J. 1978d. *H. 17 Pineapple Pests: Thrips.* Pineapple Series H: Diseases and Pests Government Printer, Pretoria, Republic of South Africa, 8pp.

Petty, G.J., (1975. Pineapple mites. *Citrus Subtropics Fruit Journal* 498-15-18.

Petty, G.J., (1978c. H.16 Pineapple Pests: Pineapple Mites. Pineapple Series II diseases and pests, Government Printer, Pretoria, Republic of South Africa, 4 pp.

Phanindrakumar, H. S., K. Jayathilakan, and T.S. Vasundhara. 1991. Factors affecting the quality of freeze dried pineapple juice powder. *Food Sci. Technol.*

Pigafetta, A. 1519. Primer viaje en torno al globo. Mexico, 1954. 141 p.

Pinon, A. 1976. Interet du desherbage chimique en plantation d'ananas. Reunion Annuelle IR FA, doc.interne. no.40.

Poignant, A. 1969. Effects of two hormones applied to pineapples during fruit formation. *Fruit* 24: 353-64.

Prakash, G.S., Reddy, B.M.C and Das, H.C. 1983. The effect of partial pinching of crown at different stages of its growth on the size and shape of fruit in Kew pineapple [Ananas *comosus* (L.) Merr.]. *Singapore j. Pri Ind*. 11 : 101-5.

Pruthi, J.S. and G. Lal. 1954. Varietal trials in canning of pineapple. *Bull. Central Food Technol. Res. Inst.*, 4: 284.

Purseglove, J.W. 1972. *Tropical Crops. Monocotyledons.* Longman, London, pp. 75-91.

Py, C. 1952. New information on fasciation in pineapples. *Fruit* 7: 342-346.

Py, C. 1955. Le C.M.U. Herbicide selectif hautement efficace pour plantation d'ananas (CMU is an efficacious selective herbicide for pineapple plantations). *Fruits* 10: 157-161.

Py, C. 1959b. La lutte contre les mauvaises herbes en plantations d'ananas (Weed control in pineapple plantations). *Fruits,* 14: 247-61, 291-99, 329-40, 369-87, 423-30.

Py, C. 1962. A comparison of urea and ammonium sulphate for fertilizing pineapples in Guinea. *Fruits* 17: 95-7.

Py, C. 1963. Les traitements hormone de floraison chez I'ananas. Personal Notes.

Py, C. 1965. Attempts to overcome water shortage–the principal limiting factor of pineapple growing in Guinea. *Fruits* 20 : 315-29.

Py, C. and Barbier, M. 1966. Summarie. Les traitements de floraison en culture d'ananas. *IFAC Bull. de information Bimensuel* No. 25: 1-8.

Py, C. and Lossois, P. 1962. Prevision de recolte en culture d'ananas. Etudes de correlations. Deuxieme partie. *Fruits* 17,75-87.

Py, C. and Silvy, A. 1954. Traitements hormones sur ananas. Methods practiques pour diriger la production. *Fruits Bull.* 11: 101-23.

Py, C. M., A. Tisseau, B. Oury, and F. Ahmada, 1957. *The Culture of Pineapples in Guinea,* Inst. Fran,ais de Recherches Fruitieres d'Outre-Mer, Paris,

Py, C., Haendler, H., Huet, R. and Silvy, A. 1956a. La fumure de l' ananas on Guinee. *Fruits* 11: 5-20.

Py, C., Lacoeucihe, J.J and Tisseau, C. 1987. The pineapple cultivation and uses. Editions. G.P. Maisonneuve and Larore, Paris, France.

Py, C., Lacoeuilhe, J.J and Teisson, C. 1987. *The Pineapple. Cultivation and Uses*. Editions G.-P. Maisonneuve, Paris, 568 pp.

Py, C., Tisseau, M.A. Qury, By and Ahmada, F. 1957a. La fumure de l' ananas en Guinee. *Fertilite*, 3: 5-25.

Py,C. 1953. Les hormones dans la culture de l'ananas. Annales Institut Franqais de Recherches Fruitieres d'Outre- Mer (IFAC) 6, 46.

Py,C. 1964. Aperecu sur le cycle de l'ananas en Martinique. *Fruits* 19,133-139.

Py,C. and Guyot. 1970. La floraison controlee de l'ananas per l'ethrel, nouveau regulateur de croissance (Fin). *Fruits*, 25: 427-445.

Py,C., Tisseau, M.A., Oury, B. and Ahmada, F. 1957. La Culture de l'ananas en Guinee-Manuel du planteur. Institut Francais de Recherches Fruitieres d'Outre-Mer (IFAC), Paris, France, 331 pp.

Rabie, E.C., Tustin, H.A. and Wesson, K.T. 2000. Inhibition of natural flowering occurring during the winter months in Queen pineapple in Kwazulu- Natal, South Africa. *Acta Horticulturae* 529,175-184.

Radha, T., Baby, L.M. and Rajamony, L. 1990. Effect of depth of trenches on high density planting of Kew pineapple. *South Indian Hort.*, 38: 183-188.

Ramirez, O.D., Gandia H. And Velez-Fortuno, H. 1972. P.R.1-67., a new pineapple selection. *Fruit varieties and Horticulture Digest.* 26, 13-15

Ramteke, R.S. 1987. Characterization and storage behaviour of aroma constituents recovered from tropical fruits. Ph.D. thesis. Mysore University, Mysore, India.

Randhawa, G.S., Dass, H.C. and Chacko, E.K. 1970. Effect of ethrel, NAA and NAD on the induction of flowering in pineapple (*Ananas comosus* L.) *Curr. Sci.* 39: 530-31.

Rao, A.N. and Wee, Y.N. 1979. Embryology of the pineapple *Ananas comosus* (L.) Merr. *The New Phytologist*, 83: 485-497.

Rao, G.G., Sharma C.B.., Chadha K.L. and Shikhamany, S.D. 1977. Pineaple–a boon for dryland agriculture. *Indian Horticulture* 21(4): 3, 30.

Rao, G.G., Sharma, C.B., Chadha, K.L. and Shikhamany, S.D. 1974. Effect of varying soil moisture regimes and nitrogen levels on plant growth, yield and quality of Kew pineapple *(Ananas comosus* (L.) Merr.). *Indian Journal Hort.* 31: 306-12.

Rashid, A.R. 1975. *Ecological Studies of* Ceratocystis paradoxa (De Seynes) *Moreau in Pineapple and Sugarcane Soils in Hawaii. Plant Pathology*, University of Hawaii, Honolulu, 74pp.

Rebolledo, M.A., Uriza, D.E.A. and Rebolledo, M.L. 2000. Rates of fruitone CPA in different applications number during day versus night to flowering inhibition in pineapple. *Acta Horticulturae* 529,185-190.

Rebolledo, Martinez,A., Uriza-Avila, D. and Aguirre-Gutierrez, L. 1997. Inhibicion de la floracion de la pina con diferentes dosis de Fruitone CPA en des -dendidades de siembra. *Acta Horticulturae* 425, 347-354.

Reddy, B.M.C. and Prakash, G.S. 1982. Standardization of optimum depth of trench of planting Kew pineapple. *Annual report*, 1982. Indian Institute of Horticultural Research, pp. 19.

Reddy, B.M.C., Das, H.C., Prakash, G.S., Subramanian, T.R. and Rao, M.H. 1983. Effect of foliar application of urea on leaf nutrient status and yield of 'Kew' pineapple. *Scientia Horticulture* 18: 225-30.

Reddy, M.N., Keim, P.S., Heinrickson, R.L. and Kezdy, F.J., (1975. primary structural analysis of sulfhydryl protease inhibitors from pineapple stem. *Journal of Biological Chemistry* 250, 1741-1750.

Reitz, R. 1983. Bromeliaceas e a malaria–*Bromelia endemica*. Flora llustrada Catarinense, Santa Catarina, 808 pp.

Rios, R. and Khan, B. 1988. Listof ethnobotanical uses of Bromeliaceae. *Journal of the Bromeliad Society*, 48: 75-87.

Rodriquez, A.R. 1932. Influence of smoke and ethylene on the fruiting of pineapples (*Ananas sativus*) Schult *Journal of Agriculture*, University of Puerto Rico 26" 5-18.

Rohrbach, K. G. and Schmitt, D.P. 1994. Part IV. Pineapple. In: Plotz, R.C., Zentmyer, G.A., Nishijima, W.T. and Rohrbach, K.G. (eds) *Compendium of Tropical Fruit Diseases*. APS Press, St Paul, Minnesota, pp. 45-55.

Rohrbach, K.G. 1980. Climate and fungal pineapple fruit disease. In: *Second Southeast Asian Symposium on Plant Diseases in the Tropics*, Bangkok, Thailand (Abstract), pp.60.

Rohrbach, K.G. 1983. Pineapple diseases and pests and their potential for spread. In: Singh, K. G. (ed.) *Exotic Plant Quarantine Pests and Procedures for Introduction of Plant Materials*. ASEAN Plant Quarantine Centre and Training Institute, Serdang, Selangor, Malaysia, pp. 145-171.

Rohrbach, K.G. and Apt, W. J. 1986. Nematode and disease problems of pineapple. *Plant Disease* 70, 81-87.

Rohrbach, K.G. and Pfeiffer, J.B. 1975. The field induction of bacterial pink disease in pineapple fruits. *Phytopathology* 65, 803-805.

Rohrbach, K.G. and Pfeiffer, J.B. 1976a. The interaction of four bacteria causing pink disease of pineapple with several pineapple cultivars. *Phytopathology* 66, 396-399.

Rohrbach, K.G. and Phillips, D.J. 1990. Postharvest diseases of pineapple. In: Paull, R.E. (ed.) *Symposium on Tropical Fruit in International Trade*. International Society for Horticultural Science, Honolulu, Hawaii, pp. 503-508.

Rohrbach, K.G. and Schenck, S. 1985. Control of Pineapple heart rot, caused by *Phytophthora parasitica and P. cinnamom*, with metalaxyl, fosetyl Al, and phosphorous acid. *Plant Disease* 69, 320-323.

Rohrbach, K.G. and Taniguchi, G. 1984. Effects of temperature, moisture and stage of inflorescence development on infection of pineapple Ananas cmomosus by Penicillium funiculosum and fusarium moniliforme var. subghutimans. *Phytopathology* 74, 995-1000.

Rohrbach, K.G., Beardsley, J.W., German, T.L., Reimer, N.J. and Sanford, W.G. 1988. Mealybug wilt, mealybugs, and ants of pineapple. *Plant Disease* 72, 558-565.

Rohrbach, K.G., Namba, R. and Taniguchi, G. 1981. Endosulfan for control of pineapple interfruitlet corking, leathery pocket and fruitlet core rot. *Phytopathology* 71, 1006.

Rohrback, K. G., and W. J. Apt, 1971. Control of *Ceratocystis paradoxa* on pineapple asexual propagative parts, *Phytopathology* 61: 1323.

Rowan, A.D., Buttle, A.J. and Barrett, A.J. 1990. The cysteine proteinases of the pineapple plant. *Biochemistry Journal* 266, 869-875.

Sakia, W.S. and Sanford, W.G. 1980. Ultrastructure of the water absorbing trichomes of pineapple *(Ananas comosus)*. *Ann. J. Bot.* 46: 7-11.

Sakimura, K. 1937. A survey of host ranges of thrips in and around Hawaiian pineapple fields. *Proceedings of the Hawaiian Entomological Society* 9, 415-427.

Sakimura, K. 1940. Evidence for the identity of the yellow-spot virus with the spotted-wilt virus experiments with the vector, *Thrips tabaci. Phytopathology* 30, 281-399.

Sakimura, K. 1963. *Frankliniella fusca,* an additional vector for the tomato spotted wilt virus with notes on *Thrips tabaci, another vector. Phytopathology* 53, 412-415.

Sakimura, K. 1966. A brief enumeration of pineapple insects in Hawaii. In: *XI Pacific Science Congress*, pp. 1-7.

Salisbury, F.B. and Ross, C.W. 1992. *Plant Physiology.* Wadsworth, Belmont, California, 682 pp

Samson, J.A. 1980. *Tropical Fruits.* Longman, London and New York.

Samuels, G. and Gandia-Diaz, H. 1960. A comparison of the yield and nutritional requirements of the red Spanish and smooth cayenne pineapples. *Proceedings Caribbean Region American Society for Horticultural Science* 4, 41-47.

Samuels, G., Landrau, P. (Jr.) and Alers S.A. 1956. Influence of phosphate fertilizers on pineapple yields. *J. Agr. Univ. Puerto Rico*, 40: 218-229.

Samuels, G., Landrau, P. (Jr.) and Oliventra, R. 1954. Response of pineapple to the application of fertilizers. *J. Agric. Univ. Puerto Rico*, 39: 1-11.

Sanches, N. F. 1999. Pragas e seu Contrôle. In : Cunha, G.A.P.d., Cabral, J.R.S. and Souza, L.F.S.d. (eds) *O abacaxizeiro: Cultivo, agroindustria e economia.* EMBRAPA-SCT, Brasilia, Brazil, pp. 307-341.

Sandhu, K.S., B.S. Bhatia, and F.C. Shukla 1985. Physicochemical changes during storage of kinnow mandarin orange and pineapple juice concentrates. *J. Food Sci. Technol.* 22(5): 342.

Sanewski, G.M., Sinclair,E., Jobin-Décor,M. and dahler,G.(1998) Preliminary studies into the effects of temperature on flower initiation of smooth cayenne in south east Queensland. In: *Abstracts, Third International Pineapple Symposium.* Pattaya, Thailand. Horticultural Research Institute, Bangkok, pp.57.

Sanford, W.G. 1962. Pineapple crop log-concept and development. *Better* Craps *With Plant Food* 46, 32-43.

Sanford, W.G. 1962. *Plant Crop Results from Recent Planting Density Trials with Smooth Cayenne.* Research Report 89, Private Document, Pineapple Research Institute of Hawaii, Honolulu.

Sanford, W.G., Gowing, D.P.,Young.H.Y. and Leeper, R.W. 1954. Toxicity to pineapple plants of biurant found in urea fertilizers from different sources. *Science,* 120: 349–350.

Sasaki, M., Kato, T. and Iida, S. 1973. Antigenic determinant common to four kinds of thiol-proteases of plant origin. *Journal of Biochemistry* 74,635-637.

Savage, C.G. and Barnett, G.B.1934. Paper mulch for pineapples. *Agr. Gaz.* N.S. Wales 45 : 335-36.

Scharrer, K. and Jung, J. 1955a. Einflub der Ernahrung auf das Verhaltnis von Kationen zn Anionen in der Pflanze. Z. PftErnahr. Dung. Bodenkd., Band 71(116):76-94.

Scharrer, K. and Jung, J. 1955b. Weitere Untersuchungen uber die Nahrstoffaufnahme und das Verhatris von Ktionen zu Anionen in der Pflanze, Z. Pfl. Ernahr. Dung., Bodenkd., Band 71(116) :97-113.

Schultes, J.A and Schultes, J.H. 1830. *Annanas sativas n.n Systema Vegetabilis* 7, 1283–1285.

Scott, C. 1992. The Effect of plant density on spring plant crop/ autumn first ratoon crop cycles; test number 718. *In: Pineapple*

Field Day Book. Pineapple Industry Farm Committee, Beerwah, Queensland, pp.37-45.

Scott, C. 1993. The effect of Artificial flower inductants; 1 Nitrogen rates on juice nitrates and yield in pineapples. In: *Pineapple Field Day Notes*. Queensland Fruit and Vegetable Growers, Beerwah, Queensland, Australia, pp. 21-29.

Scott, C. 1994. Update of nitrate trials. In: *Pineapple Field Day Notes*, Queensland Fruit and Vegetable Growers, Beerwah, Queensland, Australia, pp. 9-21.

Scott, C. 2000. The effect of molybdenum applications on the juice nitrate concentration of pineapples. In: *Pineapple Field Day Notes*. Queensland Fruit and Vegetable Growers, Beerwah, Queensland, Australia, pp. 23-27.

Scowcroft, W.R. 1984. Genetic variability in tissue culture: Impact on Germplasm Conservation and Utilization. *International Board of Plant Genetic Resources*, Rome.

Selvaray, Y. Dirakar, N.G., Subhas Chander, M., Chadha, K.L. and Melanata, K.R. 1975. Biiochemical changes associated with growth and development of pineapple fruit Kew IV. Changes in major lipid constituents. *Indian Journal of Horticulture* 32, 64-67.

Sen, S.K., 1985. Pineapple, *Fruits of India: Tropical and Subtropical* (T. K. Bose, ed.), Naya Prokash, Calcutta, p. 298.

Senanayake, Y.D.A. and Gunasena, H.P.M. 1975. A study on the influence of crown leaves on fruit growth of pineapple *Ananas comosus* (L.) Merr. Cv. Kew. *Journal of Natural Agriculture Society*, Ceylon 12, 106-114.

Serrano, F.B. 1928. Bacterial fruitlet brown rot of pineapple. *Philippine Journal of Science* 36, 271-305.

Sether, D.M. and Hu, J.S. 1999. Mealybugs and pineapple mealybug wilt associated virus are both necessary for mealybug *wilt. Phytopathology* 89: 70.

Sharma, C.B., Rao, G.G., Chadha, K.L and Shikhamany, S.D. 1971. *Annual Report of the Indian Institute of Horticultural Research*, Hessaraghatta, Bangalore, for 1972, pp.117.

Sideris, C.P, Young, H.Y. and Krauss, B.H. 1943. Effects of iron on the growth and ash constituents of *Ananas comosus* (L.) Merr. *Plant Physiology* 18, 608-632.

Sideris, C.P. 1955. Effects of sea water sprays on pineapple plants. *Phytopathology* 45, 590-594.

Sideris, C.P. and Krauss, B.H. 1936. The classification and nomenclature of groups of pineapple leaves, sections of leaves and sections of stems based on morphological and anatomical differences. *Pineapple Quarterly*, 6: 135-147.

Sideris, C.P. and Krauss, B.H. and Young Y.N. 1939. Distribution of different nitrogen fractions, sugars and other substances in various sections of the pineapple grown in soil cultures and receiving either ammonium or nitrate salts. *Plant Physiol.* 14: 227-54.

Sideris, C.P. and Young, H.Y. 1945. Effects of different amounts of potassium on growth and ash constituents of *Ananas comosus* (L.) Merr. *Plant Physiology* 20, 609-630.

Sideris, C.P. and Young, H.Y. 1946. Effects of nitrogen on growth and ash constituents of *Ananas comosus* (L.) Merr. *Plant Physiology* 21, 247-270.

Sideris, C.P. and Young, H.Y. 1951. Growth of *Ananas comosns* (L.) Merr, at different levels of mineral nutrition under greenhouse and field conditions. II. Chemical composition of the tissues at different growth intervals. *Plant Physiology* 26, 456-474.

Sideris, C.P. and Young, H.Y. 1956. Pineapple chlorosis in relation to iron and nitrogen. *Plant Physiology* 31, 211-222.

Sideris,C.P and Krauss, B.H. 1938. Growth Phenomena of pineapple fruits. Growth 2,181-196.

Sinclair, E. 1994. Borax or Solubor for Ethrel induction? In: *Pineapple Field Day Notes*. Queensland Fruit and Vegetable Growers, Beerwah, Queensland, Australia, pp. 64-66.

Sinclair, E.R. 1991. Progress in ethrel flower induction studies by QDPI/CSIRO. In: *Pineapple Field day Book*. Pineapple Industry Farm Committee, Beerwah, Queensland, pp.39-46

Sinclair, E.R. 1992b. Yield of pineapple in relation to plant size and season. *In: Pineapple Field Day Book*. Pineapple Industry Farm Committee, Beerwah, Queensland pp.23-29.

Sinclair, E.R. 1997. Weather summary for 1996/97. In: *Pineapple Field Day Book*. Pineapple Industry Farm Committee Beerwah, Queensland, p. 29.

Sinclair, E.R. 1998. Weather summary for 1996/98. In: *Pineapple Field Day Book*. Pineapple Industry Farm Committee Beerwah, Queensland, p. 72.

Sinclair, E.R. 1999. Weather summary for 1996/99. In: *Pineapple Field Day Book*. Pineapple Industry Farm Committee Beerwah, Queensland, p. 72.

Singh, D.B and B.L. Attri. 1999. Postharvest changes in pineapple var. Queen during storage. *J.Andaman Sci. Assoc.* 15(2):57-59.

Singh, D.B. 1997. Sucker production in pineapple var. 'Kew' without effecting yield and quality of fruit. *J. Indian Soc. Coastal Agric. Research.* 15(2):47-50.

Singh, D.B. and B.L. Attri. 1999.Standardization of flower induction time in pineapple var. 'kew'. *The Hort. J.* 12(2): 27-32.

Singh, D.B. and T.V.R.S. Sharma. 2002. Role of Paclobutrazol and Thiourea on Sucker induction in pineapple var. 'kew'. *India J. Hort.*, 59(4): 382-385.

Singh, D.B., A.B. Mondal. 2000. Assessment of pineapple plants developed from micropropagation instead of conventional suckering. *Tropical Science.* 40 (4) 169-73.

Singh, D.B., A.K. Bandyopadhyay. 1999. Year round pineapple production. *Success stories on transfer of technology in SAARC Countries*, pp.1-4.

Singh, D.B., T.V.R.S. Sharma and M.A. Suryanarayana. 1994. Quick multiplication of Pineapple with Crown cuttings. *J. Andaman Sci. Assoc.*, 10 (1 and 2): 92-93.

Singh, D.B., Vijai Singh and A.K. Bandyopadhyay. 2002. Growth and Development in Pineapple var. 'Kew' as influenced by Nitrogen and Phosphorus levels. *Progressive Hort.* 36(1): 44-50.

Singh, D.B., Vijai Singh, T.V.R.S. Sharma and A.K. Bandyopadhyay. 1999. Flowering and fruiting in Pineapple var. 'kew' as affected by plant growth regulators. *Indian J. Hort.* 56(3):223-228.

Singh, D.B.; T.V.R.S. Sharma and A.K. Bandyopadhyay. 1997. Efficacy of NAA for off season fruit production of pineapple var. 'Kew'. *J. Indian Soc. Of Coastal Agricultural Research*, 15(1) :61-64.

Singh, H.P. and Rameshwar, A.1976. Efficiency of calcium carbide in inducing flowering in pineapple in Malnad area of south Indian, *Indian J. Hort.* 51: 157-59.

Singh, H.P., Dass, H.C. and Ganapathy, K.M. 1978. Studies on different types and size of plating material in Kew Pineapple. From Research Report and project proposals on banana, pineapple and papaya presented at Fruit Research Workshop held at Bangalore 31-7-1978, pp. 134-5.

Singh. D.B, M.A. Suryanarayana and TVRS Sharma. 2002. Effect of NPK on growth, yield and quality of pineapple cv. Queen under rainfed conditions of Andaman. *Indian J. Hort.* 59(3): 261-265.

Singh.D.B, Vijay Singh and A.K.Bandyopadhaya. 1998. Effect of planting time and size of sucker on year round pineapple production. *J. Andaman science Assoc.* 14(1):7-15.

Sita, G.L., Sigh. R. and Iyer, C.P.A 1974. Plantlets through shoot-tip cultures in pineapple. *Curr. Sci* 43: 724-25.

Smith, L. G. 1988. Indices of physiological maturity and eating quality in Smooth Cayenne pineapple. Indices of eating quality, *Queensland J. Age Anim. Sci.* 5(2): 219.

Smith, L.B. 1939. Noteson Taxonomy of Ananas and pseudananas. *Botanical Museum Leaflet*, Harward 7: 73-81

Smith, M.K and Drew, R.A. 1990. Current applications of tissue culture in plant propagation and improvement. *Australian Journal of Plant Physiology* 17, 267-289.

Snowdon, A.L. (1990) *Color Atlas of Postharvest Diseases and Disorders of Fruits and Vegetables.* CRC Press, Boca Raton, Florida.

Soler, A. 1992a. Pineapple, CIRAD-IRFA, Paris, France, 48 pp.

Srivastava, R.P. 1963. Hunger signs in pineapple. *Fertilizer News* 8(9): 7-11.

Srivastava, S.S. 1960. Effect of foliar application of zinc on growth, fruiting behavior and quality of pineapple. *Indian J. Hort.* 26: 145-50.

Su, 1958. N.R. 1958. The response of pineapple to the application of potassium chloride. I. Plant Crop. Journal of the agricultural Association of China. 22, 27-50.

Su, N.R. 1957a. Spacing and fertilizer level as two dominant factors in the production of pineapples. *J. Agri. Assoc. China* 17: 42-67.

Su, N.R. and Haung, C.P. 1956. Effect of foliar fertilization after the initiation of floral differentiation of pineapple. *J. Agri. Assoc. China* 14: 30-37.

Su, N.R., Huang C.R. and Chow, Y.F.1956. Effect of rice straw mulching on pineapples. *Soils Fer, Taiwan* 5:42-43.

Subramanian, T.R., Chadha, K.L., Srinivasa Murthy, H.K. and Melanta, K.R. 1972. Studies on nutrition of pineapple. II. Effect of varying levels of NPK on leaf nutrient status in variety Kew in low fertility soils. Paper presented in *Third Int. Tropical and Subtropical Hort.*, Bangalore, February.

Subramanian, T.R., Chadha, K.L., Srinivasa Murthy, H.K. and Melanta, K.R. 1974. Studies on nutrition of pineapple. III. Effect of varying levels of NPK on leaf nutrient status in variety Kew in lhigh fertility soils. *Indian J. Hort.*, 31: 219-22.

Sweta Kelly, D.E. 1993. Nutritional Disorders. In: Broadley, R.H., Wassman, R.C. and Sinclair, E.R. (eds) *Pineapple Pests and Disorders*. Department of Primary Industries, Brisbane, Queensland, pp. 33-42.

Taiz, L. and Zeiger, E. 1991. *Plant Physiology*. Benjamin/Cummings, Menlo Park, California, 559 pp.

Tan, K.M. and Wee, Y.C. 1973. Influence of size of pineapple slips on plant growth, fruit weights and quality in graded and mixed plantings. *Tropical Agriculture*, Trinidad.

Tan, T.H. 1974. Effect of water on growth and nutrient uptake of pineapple. *MARDI Research Bulletin* 2, 31-49.

Tay, T.H. 1974. Effect of potassium and magnesium application on the yield and quality of pineapple. *Malaysian Agricultural Research and Development Institute Research Bulletin* 2,43-45.

Tay, T.H., (1974. Effect of water on growth and nutrient uptake of pineapple. *MARDI Research Bulletin* 2, 31-49.

Tay, T.H., (1977. Fruit ripening studies on pineapple. *MARDI Research Bulletin* 4, 29-34.

Teaotia, S.S. and Pandey, I.C. 1966. Effect of planting material on survival, growth and flowering in pineapple var. Giant Kew [*Ananas comosus* (L.) Merr.]. *Indian J. Hort.* 23: 127-30.

Teisson, C. 1972. Etude sur la floraison naturelle de l'ananas en Cote d'Ivoire. *Fruits* 27, 699-704.

Teisson, C. 1973. *Essai ombrage du fruit.Etat nutritive et qualite du fruit.* Document interne Nos. 47 and 141. Reunion Annuelle,IRFA, Montpellier.

Teisson, C. 1977. Le brunissement interne de l'ananas. Docteur es sciences naturelles these, presentee a la Faculte des Sciences de l'Universite d'Abidjan, 184 pp.

Teisson, C. 1979. Internal browning of pineapple. I. Review, II. Materials and methods (disorders). *Fruits* 34(4), 245-261.

Teisson, C. and Pineau, P. 1982. Quelques donnees sur les dermieres phases du development de l'ananas. *Fruits* 37, 741-748.

Teiwes, G and Gruneberg, F. 1963. Science and practice in the manuring of pineapples. *Green Bull. Hannover* 3:11-64.

Thamsurakul, S., Nopamornbodi, O., Charoensook, S. and Roenrungroeng, S. 2000. Increasing pineapple yield y using VA mycorrhizal fungi. *Acta Horticulturae* 529-202.

Tisseau, M.A and Tisseau, R. 1963. The setting up of an experiment on the flowering and mineral nutrition of pineapple. Fruits 18: 33-36.

Tisseau, M.A. 1959. La deficiency en cuirre et en zinc chez l' ananas:le "Crook-neck". *Fruits d'Outre Mer*. 14: 363-67.

Tisseau, R. 1982. Surmaturation interne des ananas de Cote-d'Ivoire. Le "jaune". Incidence sur la qualite gustative des fruits Commercialises en France. Reunion Annuelle IRFA, dic. Interne, No. 1.

Tisseau, R., Teisson, C., Soler, A., Huet, R. and Crochon, M. 1981. Recherches sur la qualite des ananas. Colloque sur l'Agroindustrie du 3/12/81. *Document IRFA* non publie.

Traub, H.p., Cooper, W.C. and Reece, P.C. 1940. Inducing flowering in the pineapple, *Ananas sativus. Proceedings of the American Society for Horticultural Science* 37,521 525.

Treto, E., Gonzales, A. and Gomez, J.M. 1974. Etude de differentes densites de plantation chez la variete d'ananas Espanola Roja. *Fruits* 29,279-284.

Turnbull, C.G.N., Anderson, K.L., Shorter, A.J., Nissen, R.J. and Sinclair, E.R. 1993. Ethephon and causes of flowering failure in pineapple.*Acta Horticulturae* 334,83-92.

Upadhya, M. D. and J.L. Brewbaker. 1966. Effect of gamma irradiation on the pineapple. *Hawaii Farm Science* 15(1): 8–9.

Upadhya, M.D., J.L. Brewbaker and K.W. Ching. 1966–67. Biochemical chages in gamma-irradiated pineapple. U.S. Atomic Energy Commission Div. of Isotope Development. *Ann. Rep. J.* p. 3

Van Lelyveld, L.J., and de bruyn, J.A., (1977. polyphenols, ascorbic acid and related enzyme activities associated with black heart in Cayenne pineapple fruit. *Agrochemophysica* 9, 1-6.

Van Overbeek, J. 1946. Control of flower formation and fruit size in the pineapple. *Botanical Gazette* 108, 64-73.

Van Overbeek, J. and Cruzado, H.J. 1948a. Flower formation in the pineapple plant by geotropic stimulation. *American Journal of Botany* 35: 410-412.

Van Overbeek, J. and Cruzado, H.J. 1948b. Note on flower formation in the pineapple induced by low night temperature. *Plant Physiology* 23: 282-285.

Varkey, P.A., Nair, M.N.C., George, T.E., Aipe, K.C and Leela Mathew. 1984. Growth, flowering duration and growth characters as influenced by size of propagules in pineapple (*Ananas comosus* Merr.) cv. Kew. *South Indian Hort.* 32: 327-29.

Vasconcelos, D.M. 1952. Adubacao de abacaciero. Bol. Secr. Agric. Ind. Com. Est. Pernambuco, 207.

Ventura, J.A., Maffia, L.A. and Chaves, G.M. 1981. *Fusarium moniliforme* Sheld var. *subglutinans* Wr. and Rg. (*Ananas comosus*). *Fruits*, 36: 707-710.

Villegas, V.N., Pementel, R.B., Siar, S.V. and Barile, E.B. 1995. Preliminary evaluation of promising pineapple hybrids. In: Abstracts of Second Symposium International Ananas, 20-24 February 1995. CIRAD-ISHS, Trois-Ilets, Martinique.

Waite, G.R. 1993. Pests. In: Broadly, R.H., Wassman, R.C. and Sinclair, E.R. (eds) *Pineapple Pests* and *Disorders.* Department of Primary Industries, Brisbane, Queensland, Australia.

Waithaka, J.H.G. and Puri, D.K. 1971. Recent research on pineapple in Kenya. *World Crops* 23: 190-92.

Wakasa, K. 1989. Pineapple. In: Bajaja, Y.P.S. (ed.) *Biotechnology in Agriculture and Forestry 5: Tree II.* Springer-Verlag, Berlin, pp. 13 -29.

Wakasa, K. 1989. Pineapple. In: Bajaja, Y.P.S. (ed.) *Biotechnology in Agriculture and Forestry* 5: Trees II. Springer-Verlag, Berlin, pp. 13-29.

Wardlaw, C.W., (1937. Tropical fruits and vegetables: an account of their storage and transport. *Tropical Agriculture* 24, 288-298.

Wasaka Kyo *et al.* 1978. Differentiation from in vitro culture of *Ananas comosus. Japan J. Breed* 28 (2) : 113-21.

Wasaka Kyo. 1979. Variation in plants differentiated from the tissue culture of pineapple. Nat Inst. of Agric. Sci Atahe Ibaraki 300-21. *Japan J. Breed.* 29 (1): 13-22.

Wassman, R.C. 1982. The importance of selected clones in pineapple production. In: *Pineapple Field Day Book.* Pineapple Industry farm Committee, Beerwah, Queensland, p.28.

Wassman, R.C. 1991. The high cost of inadequate fruit induction: summer forcing for spring harvest. In: *Pineapple Field Day Book.* Pineapple Industry farm Committee, Beerwah, Queensland, pp.37-38.

Waterhouse, D.F. and Norris, K.R. 1987. *Control Pacific Prospects.* Inkata Press. Melbourne, Australia.

Wayman, O., Kellems, R.O., Carpenter, J.R. and Ngugen, A.H. 1976. Potential feeding value of whole pineapple plants. *Journal of Animal Science,* 42: 1572.

Wee Y.C. 1974. The Masmerah pineapple: a New Cultivar for the Malayasian pineapple industry. *World Crops* March/April, 67-76.

Wee, Y.C. 1969. Planting density trials with *Ananas comosus* (L.) Merr. Var. Singapore Spanish. *Malays. Agric.* J. 47: 164-74.

Wee, Y.C. 1978 Natural Flowering and fruiting of the Singapore Spanish pineapple in Johore, West, West, West Malaysia. *Journal of the Singapore National Academy of Science* 7, 9-14.

Wee, Y.C. and Ng, J.C. 1968. Some observations on the effect of month of planting on the 'Singapore Spanish variety of pineapple'. *Malays. Agric. J.* 46:469-75.

Wee, Y.C. and Ng, J.C. 1970. Flower induction in pineapple cultivation. J. *Singapore Nat. Acad. Sci.* 2:69-73.

Wee, Y.C. and Ng, J.C. 1971. The effects of ethrel on the Singapore Spanish pineapple. *Malays. Pinapple* 1: 5-10.

Wee, Y.C. and Rao, A.N. 1979. *Ananas* pollen germination. Grana 18, 33-39.

Wee,Y.C. and Rao, A.N. 1977 Flowering response of Singapore Spanish pineapple to monthly applications of acetylene and NAA. *Malaysian Agricultural Journal* 51,154-166.

Wenkam. N. S. and A.P. Mol. 1967–68. Nutritional composition of irradiated fruit. II. Pineapple. U.S. Atomic Energy Commission. Division of Isotope Development. *Ann. Rep.*, p. 157.

William, D.D.F. and Fleisch, H. 1993. Historical review of pineapple breeding in Hawaii. *Acta Horticulture* 334, 67-76.

Williams, D.D.F. 1987. History and Development of fruit differentiation, growth and ripening control in pineapple. In: *Proceeding Plant Growth Society of American 14th Annual Meetings.* Plant Growth Regulations Society of American, LaGrange, Georgia, pp.413-422.

Wolf, D.E. 1953. CMu- Its use as a herbicide. Proc. 5th Annu. Calif. Weed. *Conf.*, pp 77-79.

Yamada, F., Tkahashi, N. and Murachi, T. 1976. Purification and characterization of a proteinase from pineapple fruit, fruit bromelain FA2. *Journal of Biochemistry* 79, 1223-1234.

Yamane, G.M. and Ito, O. 1969. ACP 66-329, a potential forcing agent. *Pineapple News* 17, 1-27. Private Document, Pineapple, Pineapple Research Institute of Hawaii, Honolulu.

Yamane, G.M. and Ito, O. 1970. Recent developments in Ethrel forcing. *Pineapple News* 18, 13-23. Private Document, Pineapple, Pineapple Research Institute of Hawaii, Honolulu.

Yang, S.F., and Hoffman, N.E. 1984. Ethylene biosynthesis and its regulation in higher plants. *Annual Review of Plant Physiology* 35,155-189.

Yow, Y.L. 1959. The time of maturity for the summer crop of pineapples in relation to climatic and cultural conditions. *Journal of the Agricultural Association of China* 27,26 46.

Yow, Y.L., (1972. diurnal variation in flowering response of pineapple plants to application of acetylene, naphthaleneatetic acid and beta-hydroxyethyledraine. In: *International Conference on Tropical and Subtropical Agriculture. American Society of Agriculture Engineering*, St Joseph, Missouri, p.241.

Zepeda, C. and Sagawa, Y. 1981. *In vitro* propagation of pineapple. *HortScience* 16, 495.

Zhang, J. 1992. Computer simulation of pineapple growth, development, and yield. *Ph.D. Dissertation*. University of Hawaii at Manoa, Honolulu, Hawaii.

Zhang, J. and Bartholomew, D.P. 1997. Effect of plant population density on growth and dry matter partitioning of pineapple. *Acta Horticulturae* 425: 363-376.

Zhu, J. 1996. Physiological responses of pineapple [*Ananas comosus* (L.) Merr.] to carbon dioxide enrichment, temperature and water deficit. *Ph.D. Dissertation*, University of Hawaii at Manoa, Honolulu, Hawaii

Zhu, J., Goldstein, G. and Bartholomew, D. 1999. Gas exchange and carbon isotope composition of *Ananas comosus* in response to elevated carbon dioxide and temperature. *Plant Cell and Environment* 22: 999-1007.

Zimmerman, E.C. 1948. *Insects of Hawaii.* A Manual of the Insects of the Hawaiian Islands Including Enumeration of the Species and Notes on the Origin, Distribution, Hosts, Parasites, etc. University Press of Hawaii, Honolulu, Hawaii.

Index

www.ingramcontent.com/pod-product-compliance
Lightning Source LLC
Chambersburg PA
CBHW060931240326
41458CB00139B/869